Radioactive Waste Management

Radioactive Waste Management

Edited by **Peggy Sparks**

LANRYE
INTERNATIONAL

New Jersey

Published by Clanrye International,
55 Van Reypen Street,
Jersey City, NJ 07306, USA
www.clanryeinternational.com

Radioactive Waste Management
Edited by Peggy Sparks

International Standard Book Number: 978-1-63240-436-7 (Hardback)

Contents

Preface

Radioactive waste is a topic that has currently received huge recognition due to the large amount of accumulative wastes and the increased public awareness of the hazards of such wastes. This book caters to the practice and research efforts that are lately conducted to deal with the technical problems in various radioactive waste management acts and to introduce the non-technical factors that can affect the management applications. The mutual contributions of esteem international experts have presented the science and technology of several management practices under the section disposal activities. The authors have covered the management system; presented how old management methods and radioactive accidents can affect the environment in both pre-disposal and post-disposal acts.

All of the data presented henceforth, was collaborated in the wake of recent advancements in the field. The aim of this book is to present the diversified developments from across the globe in a comprehensible manner. The opinions expressed in each chapter belong solely to the contributing authors. Their interpretations of the topics are the integral part of this book, which I have carefully compiled for a better understanding of the readers.

At the end, I would like to thank all those who dedicated their time and efforts for the successful completion of this book. I also wish to convey my gratitude towards my friends and family who supported me at every step.

Editor

Disposal Activities

Statistical Analyses of Pore Pressure Signals in Claystone During Excavation Works at the Mont Terri Underground Research Laboratory

Rachid Ababou[1], Hassane Fatmi[1],
Jean-Michel Matray[2], Christophe Nussbaum[3] and David Bailly[1,2]
[1]IMFT, Institut de Mécanique des Fluides de Toulouse
[2]IRSN, Institut de Radioprotection et de Sûreté Nucléaire
[3]SWISSTOPO, Mont Terri Consortium
[1,2]France
[3]Switzerland

1. Introduction

In many countries (such as Belgium, Germany, France, Japan, Switzerland, and United Kingdom), deep argillaceous formations are considered as potential host rocks for geological disposal of high-level and intermediate-level long-lived radioactive wastes. Some of these countries are investigating the suitability of high compacted clay-rich rocks at depths down to around 500 m below the ground surface. The general disposal concept comprises a network of drifts and tunnels linked to the surface by shafts and ramps, all artificially ventilated. Research is ongoing in Underground Research Laboratories, like the Mont Terri site in the Swiss Jura, to assess and ensure the safety of the repositories for the full decay life of the radioactive waste, i.e. the capacity of the hypothetical repository to prevent the migration of radionuclides towards the biosphere.

One of the important safety considerations regarding nuclear waste disposal repositories are damages and disturbances induced by excavation works to the rock-mass in the vicinity of the openings. These excavations can alter the initial and favorable isolation properties of the host rock. The most obvious problem is that of changes in the stress field around the tunnel during excavation that induces a fracture network, namely the Excavation Damaged Zone or EDZ, consisting mainly in unloading joints and shrinkage cracks (Bossart et al., 2002) and therefore characterized by irreversible changes. This phenomenon creates an increase in permeability of potentially several orders of magnitude higher than that of the undamaged host rock (Blümling et al., 2007). The EDZ and its raised permeability could possibly affect the performance and safety of the repository as it could act as a preferential pathway, allowing radionuclides to migrate to the surface, bypassing bentonite seals along the drifts.

However, excavation works may also contribute to reversibly disturb the geological medium in the so-called Excavation disturbed Zone (EdZ). This term (EdZ) designates a zone that is thought to be disturbed by elastic hydro-mechanical variations of pressure and stress, but without irreversible changes in flow and transport properties. Within the EdZ, there are no negative effects on the long-term safety, but short-term effects may exist as the response to hydro-mechanical coupling. For instance, mine-by test experiments performed during excavation works often show an increase in pore pressures in the vicinity of the newly excavated gallery. This increase may exceed 1 MPa (10 bars), for instance during excavation of Ga03 in the Tournemire URL (Massmann, 2009). This pore pressure increase appears either instantaneous or delayed depending on the position of the measurement sections with respect to the excavation front.

In short, the objective of this text is to study the isolation properties of a geologic repository against the migration of radionuclides through the host rock and eventually towards the biosphere. In this context, the hydro-geologic properties of the host rock must be characterized at different scales (e.g., near field vs. far field).

The permeability is a very important property in terms of isolation, but in this work, we focus on novel methods for identifying two other related properties: (i) specific storativity (as the elastic response to earth tides), and (ii) the effective dynamic porosity (using the barometric effect). Indeed, the analysis of pore pressure and atmospheric pressure time series makes it possible in some cases to determine hydraulic parameters such as specific storage coefficient and effective porosity using simplified groundwater flow and compressibility models (e.g., Marsaud et al. 1993). These models have already been applied to obtain a preliminary characterization of the compacted clay rock of the Tournemire URL in Aveyron, France (Fatmi et al. 2005).

More precisely, our main goal in this text is to characterize the above mentioned hydraulic and mechanical properties of the clay rock from analyses of measured atmospheric and pore pressure signals, and also, to study the response of the clay rock's "Excavation disturbed Zone" ("EdZ") during the excavation of galleries.

The data analysis techniques rely strongly on statistical concepts (theory of random processes) and on the decomposition of signals on various bases:

- Harmonic basis (sine, cosine): Fourier spectrum S(f) vs. frequency (f).
- Wavelet multiresolution bases (Daubechies): wavelet components Cj(t) at different time scales "j".

These techniques are used for interpreting the proposed hydromechanical models, in order to identify the hydrogeologic properties of the claystone from measured signals (possibly disturbed by excavation works):

- Wavelet decomposition of pore pressure signals (selection of semi-diurnal earth tide component), and statistical envelope analysis (effect of moving excavation front): earth tide effect model → specific storativity;
- Spectral cross-analysis (spectral gain) of air pressure vs. pore pressure signal: barometric effect model → porosity.

The study presented in this text is entirely based on data and time series collected at the excavated geologic claystone site of the Mont Terri URL in the Swiss Jura (Mont Terri Project - international consortium), and we focus on claystone properties at "near field" scales, within meters or tens of meters from excavated galleries at the Mont Terri site. Similar techniques are currently being developed for other datasets collected at the Tournemire URL (Aveyron, France) operated by the "Institut de Radioprotection et de Sûreté Nucléaire" (France).

Statistical Analyses of Pore Pressure Signals in Claystone During Excavation Works at the Mont
Terri Underground Research Laboratory

5

In this work, in order to achieve the above mentioned objectives, we use pressure signals obtained over a period of ten years at Mont Terri (1996/2005), from which records as long as *one year or more* will be analysed in this text (see *Section 4*). But, to analyse the effects of a moving excavation front, we also focus on a period of 5 months of *"syn-excavation"* period for gallery "Ga98", and on a much shorter period of fast pressure changes lasting only a couple of weeks (see *Section 5*). In total, we study both long and short time scale records, from years to weeks.

Some results of pressure signal pre-processing and long time scale analyses were presented in *Fatmi et al.* (2007, 2008) at an early stage of this study. The present text provides a significant update of the results of long time scale analyses, and presents also new types of results on the analysis of excavation effects on pore pressures at very short time scales (weeks).

Accordingly, the outline of this chapter is as follows.

Section 2 describes the hydro-geologic site and the pore pressure measurements at the Mont Terri site (in the framework of experiment LP14).

Section 3 is devoted to the mathematical and statistical methods of signal pre-processing (e.g. reconstruction of missing data) and of signal analysis (time-lag correlation functions, frequency spectra and cross-spectra, wavelets, envelopes). For hydro-geologic interpretation of pressure signals, the methods involve, not only statistical treatments, but also simplified elastic hydro-mechanical models: the concept of a specific storativity coefficient "Ss" (in relation to earth tides); and the barometric efficiency (involving both "Ss" and porosity "Φ") in relation to air pressure fluctuations. These models are briefly described at the end of the section (*Box N°2*) and their application is developed in the remaining sections of the text (see below).

Section 4 develops the implementation of "long time scale" statistical analyses and hydro-geologic interpretations of signals, in the case where the objective is to obtain a global estimate of the two hydro-geologic coefficients (Ss,Φ) in two steps, first "Ss" from semi-diurnal earth tide effects using multi-resolution wavelet analysis, and secondly "Φ" using the barometric spectral gain at diurnal frequency.

Section 5 develops, in contrast, a somewhat more sophisticated implementation over "short time scales" with a strong non stationary trend in the pore pressure signal. This occurs in the case of a moving excavation front passing nearby the pressure sensor. The objective is to identify the effect of the evolving "EdZ" at the location of the pressure sensor, and in particular, its effect on the evolution of Ss(t). The time scale of analysis is typically a couple of weeks, and a new tool has been developed (based on statistical envelopes) in order to identify the modulation in time of the amplitude of earth tide fluctuations.

Section 6 provides a summary of the methods and results, and suggests possible modifications and extensions of this work concerning statistical methods (multi-cross analyses) and hydro-geologic interpretations (enhanced hydro-mechanical models and inverse problems): see the "outlook" in *Sub-Section 6.2*.

The text is completed by an appendix (for abbreviations and symbols), and by a list of references.

2. Hydrogeologic setting and pressure signal measurements

2.1 Hydrogeologic setting at Mont Terri (Opalinus clay rock)

The Mont Terri Rock Laboratory was excavated within the "Opalinus Clay" of the Mont-Terri anticline, which has a very low permeability. The geological cross section shown in

Figure 1 shows the asymmetrical, dome-shaped fold of the Mont Terri anticline, where sediments of Triassic and Jurassic age are exposed.

The Opalinus clay is about 180 million years old (Mesozoic age, Jurassic period, Lower Dogger, Aalenian). It is a shallow marine mud deposit, extending over wide areas in Central Europe. It is underlain by the Jurensis Marls (Toarcian) and overlain by the "Lower Dogger" limestones (Bajocian). Three facies have been identified within this clay rock: sandy, carbonate-rich sandy, and shaly. The latter facies is studied as an analogue for hypothetical disposal of radioactive waste in a naturally isolating geologic formation (*the Mont Terri site is a research site only, not destined to host nuclear wastes or any radioactive products whatsoever*). For more details on geology and hydrogeology of the site: see *Fatmi et al. (2008), Fatmi (2009)* and *Schaeren and Norbert (1989)*. See also, more recently, *Nussbaum et al. (2011)*.

We emphasize again that the Opalinus clay rock at Mont Terri has very low permeability or hydraulic conductivity: K ≈ 10^{-13} to 5 10^{-13} m/s (1 E-13 to 5 E-13 m/s), ± 1 order of magnitude, in the intact zone outside the *Excavated Damaged Zone*. This clay rock also displays a significant self-sealing capacity due to a high proportion of swelling minerals (about 10%). In contact with water, it swells and tends to seal the cracks and fractures that occurred after excavations (galleries, niches). This self-sealing reduces permeability and participates in restoring initial rock properties.

These factors, among others, explain the choice of this type of clay rock formation by the Swiss authorities - and by the international Mont Terri consortium - as a potential host rock for high level radioactive wastes. The Mont Terri URL (URL stands for Underground Research Laboratory) is managed by SWISSTOPO. The Mont Terri consortium includes Swiss, French, British, Canadian, Spanish, Belgian, U.S., Japanese, and other nations' public authorities and their national research institutions.

Other types of geological formations were also studied by the Swiss authorities, including a crystalline rock (granite) at the Grimsel site in the Swiss Alps (e.g., FEBEX drift *in situ* experiment). Thermo-Hydro-Mechanical upscaling and modelling of the FEBEX drift *in situ* experiment was studied in *Cañamón's* thesis *(2006)*, later published as *Cañamón (2009)*. These references also include statistical signal analyses of the FEBEX thermo-migration laboratory experiment, or "mock-up" experiment (see also *Cañamón et al. (2004)*. The in situ and mock-up FEBEX experiments were performed under funding by the Spanish nuclear authorities *ENRESA*.

In the remainder of this text, we focus exclusively on the analysis of pressure signals collected *in situ* in Mont Terri's Opalinus Clay rock site.

2.2 Pressure signal measurements and pre-processing (Mont Terri BPP-1 borehole)

The pressure signals used in this work were obtained over a period of 10 years with half-hourly time steps (on average), at the Mont Terri URL, in the framework of the LP14 project (*Long term Pressure monitoring project - phase 14*). Concerning Mont Terri's LP project, see *Marschall et al. (2009)*. The BPP-1 borehole (101 mm diameter, and 20 m length) was dug in the year 1996, from niche PP located in Mont Terri's "exploration gallery". The measurement sections or sensors (PP-1 and PP-2) are located at the end of the BPP-1 borehole, outside the excavation damaged zone (EDZ) of the exploration gallery, in the silty-shaly facies of the Opalinus clay.

Statistical Analyses of Pore Pressure Signals in Claystone During Excavation Works at the Mont
Terri Underground Research Laboratory

7

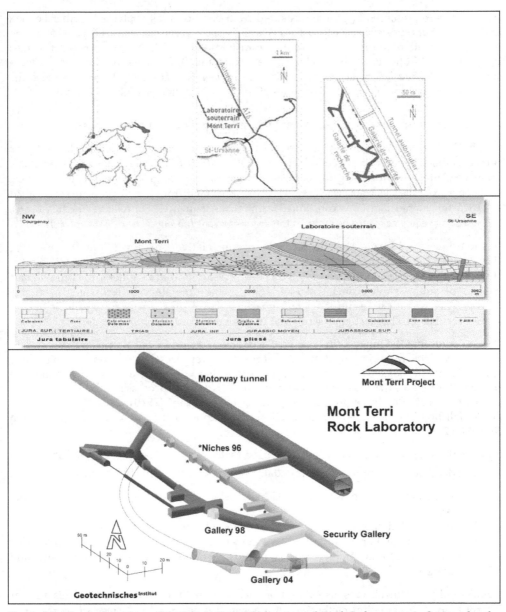

Fig. 1. **(a) Above.** Location of the Mont Terri Underground Rock Laboratory in Switzerland,
with its main experimental galleries. **(b) Middle:** Geologic cross-section of the folded Jura
showing the Mont Terri anticline and the 4 km long motorway tunnel giving access to the
Opalinus Clay rock (Freivogel M. 2001, Bâle University, after Schaeren G. & Norbert J. 1989:
Tunnels du Mont Terri et du Mont Russelin. La traversée des "roches à risques": marnes et
marnes à anhydrite. *Soc. Suisse Ing. Arch.*, Doc. SIA D 037, 19–24). **(c) Below:** Schematic 3D
perspective view of some of the Mont Terri galleries and the motorway tunnel.

Absolute pore pressures P(t) were measured from boreholes (as explained further below), and atmospheric pressure $P_{AIR}(t)$ or $P_{ATM}(t)$ was measured underground in the exploration gallery. Absolute pore pressures were then transformed into relative pressures, defined as: $P_{REL}(t) = P(t)-P_{ATM}(t)$. For more technical details on the BPP-1 borehole and its pressure sensors, see *Figure 3* further below, and see *Thury and Bossart (1999)*. Finally, see also the previous *Figure 1* for 3D perspective views of the tunnels, galleries, and boreholes.

2.2.1 Signal pre-processing, record lengths, missing data and other issues

The BPP-1 borehole was initially selected for this study because it provided the longest pore pressure times series over a period of about 10 years (from 17/12/1996 to 30/06/2005). However, after pre-processing for outliers, data gaps, and variable steps, the 10 year long sequences from the PP1 and PP2 pore pressure sensors (and from the air pressure sensor) had to be fractioned into shorter "reconstituted" contiguous sequences. For relative pressure $P_{REL}(t)$, the resulting sequences were roughly on the order of a year (*no longer than 1.5 year at best*).

It should be emphasized that, in order to obtain relative pressure sequences ($P_{REL}(t) = P(t)-P_{ATM}(t)$), the pore pressure P(t) and air pressure $P_{ATM}(t)$ had to be processed "jointly" (*Fatmi et al. 2008*). Joint pre-processing of pore pressure and atmospheric pressure signals get rid of outlier values, data gaps, and/or variable time steps in *both* signals, but the consequence is that the resulting "clean signal" of relative pressure is only available in sub-sequences that are shorter than the original 10 years of recorded data.

It was found that the longest un-processed contiguous sequence of relative pressure $P_{REL}(t)$, directly available from the raw data without need for preprocessing, was only 1 month long (subsequence 02/08/2002 - 04/09/2002). Pre-processing was necessary in order to increase the length of the $P_{REL}(t)$ records beyond one month. After pre-processing, the longest $P_{REL}(t)$ record for sensor PP1 has a duration of about 14 to 15 months (29/01/2004-12/04/2005), or 20736 half-hourly time steps. This longer, partially reconstructed signal, will be analysed in *Section 4.2*.

Statistical tests, involving recursions, were used to detect possible outliers (transformed into missing data). In the presence of missing data, for partial reconstitutionS, we used:

-   decompositions into trends and residuals using moving average filters (and other filters),
-   linear interpolation techniques (usually only for relatively short data gaps),
-   a bidirectional first order autoregressive model (going forward and backward in time)
-   statistical tests and validations of reconstituted series based on moments and covariances

Time step homogenization is also a kind of partial reconstitution. We used for this a linear interpolation/extrapolation method, with a constant time step "Δto" selected by the user. Here, all processed signals have, after "homogenization", a half-hourly time step (Δto = 30 mn). This choice is in agreement with the nominal acquisition step (also the most frequently encountered time step in the collected data).

Figure 2 shows a simplified flow chart presenting these Pre-Processing tasks and methods. For more details on pre-processing methods, see *Fatmi (2009, Chap.3)* and *Fatmi et al. (2007, 2008)*.

Statistical Analyses of Pore Pressure Signals in Claystone During Excavation Works at the Mont
Terri Underground Research Laboratory

9

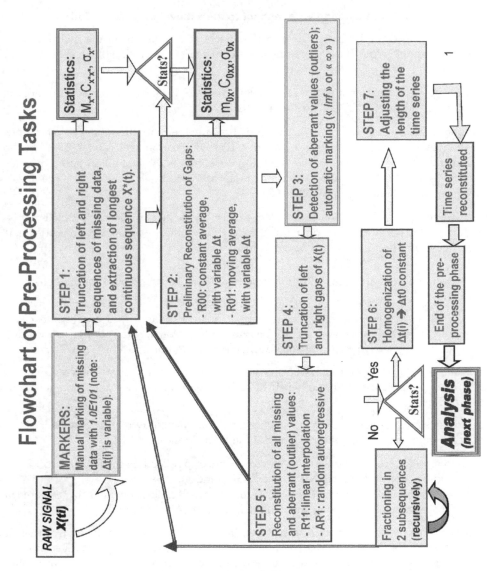

Fig. 2. Flowchart of Pre-Processing tasks and methods (programmed and implemented as a
MATLAB TOOLBOX). The pre-processing tasks contain some iterations and recursions,
particularly when computing statistics to detect outliers or to fill data gaps. Only some of
the iterative procedures are explicitly shown here. The symbols (Mx*, Cxx*, σx*) and (mox,
Coxx, σox) represent estimates of the mean, the covariance function, and the standard
deviation of the signal at various stage of pre-processing. The symbol Δto is the constant
time step chosen by the user for time step homogenisation. At the end of the pre-processing
stage, the processed (transformed) signal is ready for statistical analyses. The processed
signal may be shorter than the initial "raw" signal.

2.2.2 Pressure sensors location with respect to the excavated Gallery 1998

Since we are especially interested here in the effects of excavation on the hydrogeologic properties of the clay rock (EDZ), let us focus here on a brief description of the excavation of gallery Ga98.

The Ga98 gallery was dug at Mont Terri during the year 1998, over a period of 5 months (more precisely, from 17/11/1997 to 23/04/1998).

Recall that the main goal of this work is to study not only the properties of the claystone, but also, their evolution in response to excavation (Excavation Damaged Zone). This will be done, for the case of gallery Ga98, by analysing the effects of excavation on pore pressures collected at sensors PP1 and PP2, knowing that they are distant only a few meters (4.4 m and 4.5 m) from the excavated gallery Ga98 (*Figure 3*).

Thus, due to their close location to the 1998 gallery, the sensors of borehole BPP-1 can be used to monitor pore pressures during the excavation of the 1998 gallery: we will exploit this particular configuration in this paper, as in a "*mine-by test*".

2.2.3 Note on related analyses of other signals at Mont Terri and at other clay sites

While we focus in this work on air and pore pressures at Mont Terri, for the sake of completeness, we present in this section a brief note on other signal analyses: e.g., air temperature, relative humidity, and crack aperture at Mont Terri, and also, signals recorded at other URL sites in clay rock (Bure and Tournemire, in addition to Mont Terri). Thus:

The Mont Terri "BPP-1" pore pressure signals were compared and cross-analysed with air pressure, air temperature, and relative air humidity measured in the exploration gallery of the Mont Terri URL.

In addition, similar signals were analysed and cross-analysed at two other clay rock URL's (Underground Research Laboratories) in addition to the Mont Terri URL:

- the clay rock URL of Tournemire (Aveyron, France), operated by the French "Institut de Radioprotection et de Sûreté Nucléaire" (IRSN), and
- the clay rock URL of Bure (Meuse / Haute Marne, France), operated by the French national agency on the management of nuclear waste (ANDRA).

Concerning these sites (Tournemire and Bure), the reader is referred to the Ph.D. thesis of *Fatmi (2009)*, available on line. Also, it should be noted that despite their similarities, the three clay rock sites have different tectonic, structural and hydro-mechanical characteristics (e.g., see *Nussbaum et al. 2007*, comparing Bure and Mont Terri).

- Furthermore, crack aperture time series measurements were also developed at the Mont Terri site (Cyclic Deformation phase 14 project "CD14"). The crack displacement signal was statistically analysed. The air temperature, pressure, and relative humidity were also measured and cross-analysed together with crack displacements. Some of the statistical methods used in the present study (cf. *Section 3*) were also applied to the "cross-analysis" of crack aperture response to hydro-meteorological fluctuations at various time scales. The results of these crack aperture investigations are reported in *Möri et al. (2012, in press at the time of this writing)*.

Statistical Analyses of Pore Pressure Signals in Claystone During Excavation Works at the Mont
Terri Underground Research Laboratory

11

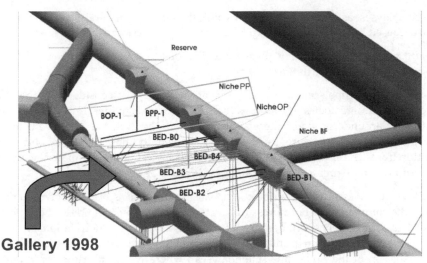

Fig. 3. Tunnels, galleries, and boreholes at the Mont Terri URL, showing Gallery 1998 ("Ga98"), and borehole BPP-1 dug from the PP niche. This niche is located in the reconnaissance gallery ("exploration gallery") parallel to the main road tunnel (dark grey). The pore pressure sensors used in this study are located near the end of borehole BPP-1 (lower left corner of blue rectangular frame), only 4 m away from Ga98. Their response during excavation of Ga98 is studied in this work.

2.3 Previous results and scope of statistical analyses this work (long and short time scales analyses of pore pressures)

The general objective of this work, in the context of radioactive waste disposal in geologic clay sites, was to exploit pressure measurements in order to quantify the phenomena affecting pore pressure (earth tides, barometric pressures,…), and *in fine*, to estimate the hydro-geologic or hydro-mechanical properties of the clay rock : elastic compressibility and/or specific storativity (Ss), effective elastic porosity(Φ), ….

Some pore pressure analyses methods and results (*Fatmi et al. 2007, 2008*), were previously developed for characterizing clay rock properties in the *undisturbed state*. These analyses used long time scale records (as long as possible), in the absence of excavation works, or prior to excavation works. The resulting estimates of Ss and Φ from pressure signals in the selected borehole at the Mont Terri URL compared favorably with those estimated by other methods, such as pulse tests on shorter time scales. Similar analyses were also carried out for other clay rock URL's: Tournemire and Bure (*Fatmi 2009*). In the present work (***Sections 4 and 5***), we provide a synthetic, updated, and upgraded presentation of the above-mentioned results, that is:

- statistical analyses and interpretations of *long pressure records*, leading to *global* estimates of average clay rock properties (Ss and Φ), but also…
- statistical analyses and interpretations of *short non stationary pressure records*, focusing on the effects of excavation works on pore pressures, and leading to estimates of *evolutionary* and *local* clay rock properties, Ss(t) and Φ(t).

All these analyses, as mentioned above, exploit the effects of natural perturbations of pore pressure, caused by hydro-meteorological as well as geophysical oscillations:

- *hydrometeorology*: atmospheric pressure oscillations;
- *planetary geophysics*: earth tides.

Thus, in the short time scale analyses (excavation works) we consider also the effects of changes in hydro-geological properties due to human activities in the URL site. The excavation of an underground gallery serves as a *"mine by"* test. The objective is to monitor the response of pore pressure (measured at a fixed location) to the evolution of the excavation front. The effect of the excavation (if any) should be superimposed on the natural perturbations of pore pressure (atmospheric and earth tide effects).

We will see that it is possible to detect, statistically, the effects of the evolving excavation front on the component of pore pressure fluctuations due to earth tides. The time scale of analysis here is relatively short, because we focus on a particular "non stationary" event, the approach of the excavation front close enough to the pressure sensor(s). This is to be distinguished from the "long time sale analyses", where we try to use records that are as long as possible, and not too much influenced by local trends.

Along these lines, for short time scale "excavation effects", our main objective is as follows:

1. to analyse the effects of drift excavation on pore pressures during excavation (syn-excavation) [1], and if possible, to analyse this effect continuously as a function of time (statistically, this boils down to analysing a signal with strongly non stationary fluctuations).

In the case at hand, we focus on the passage of an excavation front of gallery GA98 at Mont Terri (gallery 1998), and we examine closely its possible influence on the pore pressures measured in pressure sensors (PP1, PP2) of the BPP-1 borehole (*here, we will narrow this down to the single pressure sensor PP1, closest to the excavation front*). The specific objectives are, essentially:

- to quantify the evolution of elastic specific storativity $Ss(t)$ as a function of time, during the different phases (pre-, syn-, post-excavation phases), and partiuclarly, the maximum change of $Ss(t)$ during the syn-excavation phase;
- to interpret the evolution of $Ss(t)$ and to compare it with the distance $D(t)$ of the moving excavation front (and the moving "EdZ") relative to the location of the fixed pressure sensor.

3. Methods for statistical analyses and interpretations of pressure signals

Pre-processing procedures were described in the previous introductory sections (see flowchart in *Figure 2*). We now focus on methods of statistical analyses, and also, on the simplified hydro-mechanical "models" to be used for hydro-geologic interpretation of pore & air pressure analyses.

---

[1] More generally, for comparison purposes, we have also analysed pre- post-excavation phases; however the results will only be briefly mentioned here for lack of space (see *Fatmi 2009* for details).

Statistical Analyses of Pore Pressure Signals in Claystone During Excavation Works at the Mont
Terri Underground Research Laboratory

13

3.1 Overview of methods of statistical analysis (and interpretation)

After the preliminary stage of signal pre-processing, we are left with several records of contiguous time series with constant time steps (some of these signals have been pre-processed to some extent, and possibly, they have been partially reconstituted). These pre-processed pressure time series are then statistically analysed in order to characterize the hydraulic and hydro-mechanical behaviour of the porous clay rock formation (Ss, Φ). To achieve this, the following signal analysis methods were used and applied to Mont Terri signals [2]:

i.   Auto-correlation and cross-correlation analyses of pressure signals versus time lag, and also (closely related), deconvolution or identification of input/output transfer function, e.g., between atmospheric pressure and relative pore pressure.

ii. Fourier spectral analyses in the frequency domain, based on Fourier transforms of the correlation functions. This leads to identification of dominant frequencies, and also, cross-spectrum analysis and identification of the spectral gain between two signals (such as atmospheric pressure and relative pressure).

iii. Orthogonal multi-resolution wavelet analysis was used to decompose the signals in the time/scale domain (two-parameter domain). Thus, for any selected dyadic time scale (e.g., for the semi-diurnal time scale of earth tides), we isolated and extracted the corresponding "wavelet component" of the original signal. Also, we used wavelets in some cases as a filter, decomposing the original signal into a filtered quantity (the wavelet "approximation") and its residual signal (the wavelet residual) at any given dyadic time scale (the scale of the filter).

iv. Estimation of the statistical envelope of a modulated random signal based on Cramer-Leadbetter statistical envelope theory, implemented with the Hilbert transform. This technique is used to characterize the modulation, in time, of the signal extracted from semi-diurnal wavelet component analysis of pore pressure during excavation.

Each of the above-mentioned techniques was used at one stage or another in our study of Mont Terri pressure signals on various time scales.

The correlation and cross-correlation function analyses versus time lag (I), as well as the corresponding Fourier spectral analyses in frequency space (II), are both based on the theory of random processes: the reader is referred to *Bras et al. (1985); Box et al. (1976); Papoulis et al. (2002), Yaglom (1987), Yevjevich (1972).* Concerning spectral estimation and analysis (II), see also: Blackman & Tukey (1958), Max (1980), Max et al. (1996), Priestley (1981), Vanmarcke (1983). Time domain deconvolution and input/output analyses, causal or non-causal, linear or non linear, are presented in *Labat et al. (1999c; 2000a)* for the case of rainfall-runoff signals.

Multi-resolution wavelet analysis (III) is an orthogonal representation of signals in "scale-time" space (a two-parameter space). The mathematical theory is presented in *Mallat (1989).* See *Labat, Ababou et al. (1999a,b; 2000b; 2002)* concerning applications of wavelets to rainfall/runoff signals in karst hydrology. In multi-resolution wavelet analyses, "time" is

---

[2] The reader interested mostly in the applications, and not in the theoretical signal analysis methods, may jump at first reading to *Sections 4-5*, where the methods are implemented for Mont Terri pressure data.

discrete in the usual way (t(n) = n×Δt), but wavelet "scales" are dyadic (powers of 2 : 2^m×Δt). Wavelet "scales" are *analogous* to inverse frequencies, but still, they should not be confused with Fourier frequencies.

The theoretical tools III (wavelets) and IV (envelopes) will be combined and used together in *Section 5 (5.2+5.3)*. The goal is to assess, on a short time scale of a few weeks, the influence of excavation on the evolution of pore pressure and, finally, of clay rock properties in the "EdZ".

The mathematical and statistical methods cited above have well known theoretical bases, however, they can be implemented in different ways in practice. For this study, we have devised an integrated set of programs, written in MATLAB and organized in "TOOLBOXES" as follows:

•   Signal Pre-Processing TOOLBOX, and
•   Signal Analyses TOOLBOX.

The two TOOLBOXES are described in *Fatmi 2009's* PhD thesis, available on line.[3]

It should be noted that some of the methods and techniques were specifically designed or adapted by us for this study, e.g.: estimations of cross-spectra and spectral gains; identification of time domain transfer functions; filtering techniques (comparisons between moving averages and wavelet approximations); and also, an original implementation of the *Cramer-Leadbetter* envelope, performed with the *Hilbert Transform*, and applied to a selected wavelet component.

Other auxiliary studies not reported here include the design of a number of synthetic tests to validate statistical methods (cross-correlations, cross-spectra, envelopes, comparison of filtering techniques, etc). For details on some of these tests, the reader is referred to *Fatmi (2009)*.

Figure 4 shows a simplified flowchart presenting some of the signal analysis methods used in this work. These methods were programmed in a MATLAB TOOLBOX for SIGNAL ANALYSES (distinct from the PRE-PROCESSING TOOLBOX described in *Figure 2*).

3.2 Univariate (auto) correlation functions, bivariate (cross) correlation functions, and temporal transfer functions

Correlation function analysis is a way to study the temporal structure of the signals, in the time lag domain. The univariate analysis of a single signal leads to a study of its auto-correlation function. The bivariate analysis of two signals jointly leads to both auto- and cross-correlation functions.

---

[3] These MATLAB TOOLBOXES were developed mainly by H. Fatmi, supervised by R. Ababou, with the aid of the other co-authors of this chapter and of other researchers and collaborators, including: A. Mangin, senior researcher at the Moulis CNRS laboratory (who initiated several of the signal analysis procedures); doctoral students Y. Wang and K. Alastal (who tested spectral and wavelet applications at the Institute of Fluid Mechanics); and intern students A. Mallet, A. Moulia and Ch. Joly who participated in various validation tests using synthetic signals.

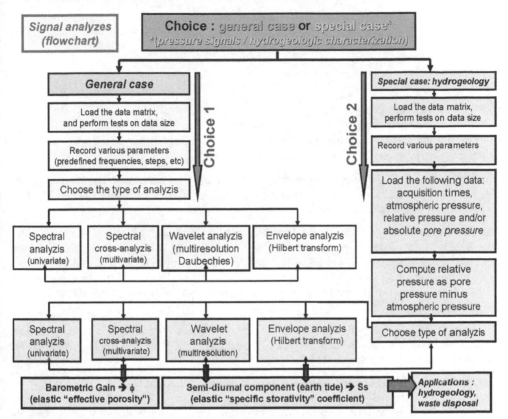

Fig. 4. Simplified flowchart presenting some of the signal analysis methods used in this work. These methods were programmed in a MATLAB TOOLBOX for SIGNAL ANALYSES.

More precisely, the temporal structure of a signal $X(t)$, or of two signals $X(t)$ and $Y(t)$ jointly, "is captured" by a temporal structure function that depends on two points in time $(t, t+\tau)$. The stationarity hypothesis simplifies this analysis by assuming that the structure function depends only on the time lag (τ). In addition, the ergodicity hypothesis (which is notoriously hard to validate in practice) further assumes that the time average of the signal is equivalent to the ensemble mean (mathematical expectation) as the record length becomes infinitely long.

Thus, for two statistically stationary and ergodic processes, $X(t)$ and $Y(t)$, the cross-covariance function is supposed to depend only on the time lag (τ) between $X(t)$ and $Y(t+\tau)$ (and it does not depend on "t"). The cross-covariance Cxy (and the cross-correlation Rxy) can be estimated as follows, from a single data record of the two processes $(X(t),Y(t))$[4]:

---

[4] The auto-covariance functions $Cxx(\tau)$ and $Cyy(\tau)$ are defined in a similar way; for instance, $Cxx(\tau)$ can be obtained simply by replacing XY by XX in the cross-covariance function $Cxy(\tau)$. The auto-correlation function $Rxx(\tau)$ is simply $Cxx(\tau)/Cxx(0)$ where $Cxx(0) = Var(X(t)) = \sigma_x^2$.

$$C_{XY}(j) = \frac{1}{N} \sum_{i=1}^{i=N-j} \left(X(t_i) - \bar{X} \right) \left(Y(t_{i+j}) - \bar{Y} \right) \text{ and } R_{XY}(j) = \frac{C_{XY}(j)}{\sigma_X . \sigma_Y} \tag{1}$$

In this equation:

- "i" is the discrete time (t(i))
- "j" is the discrete time lag (τ(j))
- $R_{XY}(j)$ is the cross-correlation (function of discrete time lag "j"),
- $C_{XY}(j)$ is the cross-covariance (function of discrete time lag "j"),
- N: total number of data in the observed time series.

One also defines "M", the maximum time lag to be retained for correlation function analyses. The discrete time lag index "j" runs through j = $\pm$\{0,1,2,…, M\}. It is necessary to keep M $\leq$ N-1 (in practice M $\leq$ N/3 or much less).

A sub-sampling step k_0 can be defined, with k_0 = 1 or 2 or 3 or … M (take k_0 = 1 if "no sub-sampling"). The time lag in physical time units is: τ_j = j.Δt (if k_0 = 1) or more generally τ_j = j.k_0.Δt ($k_0 \geq$ 1) where ko is the sub-sampling step. In our case, we choose the obvious, i.e., the smallest sub-sampling step (k_0=1) in order to sweep all the samples of the time series. For very long series or unduly small time steps, sub-sampling ($k_0 \geq$ 2) may be advisable, but this is not our case here.

The above *Eq. 1* gives a *biased* estimate of the cross-covariance function Cxy(τ). To obtain an *unbiased* estimate, the denominator (N) should be replaced by (N-j-1) in the above estimator, although some authors also use (N-j).

Finally, it is noted that Cxy(τ) is directly related to (but not equal to) the temporal transfer function Fxy(τ) between processes X(t) and Y(t). See *Labat et al. (1999c; 2000a)* for a detailed analysis and discussion of temporal transfer functions in hydrogeology & hydrology (*Instantaneous Unit Hydrographs*). Also, it is interesting to note that the transfer function Fxy(τ) can be interpreted like the "slope" of a generalized linear regression between the two signals X(t) and Y(t). The identification of this transfer function in time is an inverse problem; it is also a particular type of optimal estimation problem, also known as the "deconvolution problem".

3.3 Spectral and cross-spectral analyses, frequency gain

Fourier spectral analysis decomposes the signal (or its structure function) into periodic functions, expressed in the frequency domain using the Fourier Transform: e.g. the Fast Fourier Transform (FFT) implemented in MATLAB and many other software's.

More precisely, univariate spectral analysis consists in studying the frequency spectrum of only one signal (its auto-spectrum $S_{XX}(f)$), while bivariate or cross-spectral analysis, treats simultaneously two signals, leading to the frequency cross-spectrum $S_{XY}(f)$ of the two signals.

Other related quantities are the co-spectrum and the phase spectrum (*not used here*), and the frequency gain (*used below*). The frequency gain can be interpreted as a spectral transfer function, particularly in the case where one of the two signals (X(t)) is the "cause" of the other signal (Y(t)). This frequency gain is (in a sense) the spectral version of the temporal transfer function between the "input" X(t) and the "output" Y(t).

The spectrum $S_{XX}(f)$ of a signal X(t) can be defined and estimated in various ways (continuous/discrete time, infinite/finite time window, with/without filtering). Below, the spectrum of X(t) is expressed as a function of frequency, for the case of a discrete time process, within a limited time window, and with filtering…

3.3.1 Estimation of the Fourier spectrum (Wiener-Kinchine method, filtered)

The Fourier auto-spectrum of signal X(t) is estimated here as the Fourier transform of the correlation function Rxx(τ), based on the *Wiener-Khinchine theorem*, valid for stationary random processes. The final expression proposed here includes, also, a time-lag filter "D(τ)".

The dimensional estimated spectrum S(f) is then of the form:

$$S_{XX}(f_i) = 2\Delta t \left[C_{XX}(0) + 2\sum_{j=1}^{M} D_j.C_{XX}(j).\cos\left(2\pi\, j\Delta t\, f_i\right) \right] \qquad (2a)$$

and the reduced dimensionless spectrum s(f) is of the form:

$$s_{XX}(\hat{f}_i) = 2\left[1 + 2\sum_{j=1}^{M} D_j.R_{XX}(j).\cos\left(2\pi j\, \hat{f}_i\right) \right] \qquad (2b)$$

where:

$$R_{XX}(j) = C_{XX}(j)/\sigma_{XX}^2 \,,\, (R_{XX}(0) = 1)\,. \qquad (3a)$$

$C_{XX}(\tau)$ is the auto-covariance function and $R_{XX}(\tau)$ is the auto-correlation function, or normalized covariance function. Similarly, $S_{XX}(f)$ is the auto-spectrum, and "s_{XX}" is the reduced dimensionless spectrum (the reduced spectrum is normalized by the variance of the signal and expressed in terms of dimensionless frequency $\hat{f}_i$). Furthermore, in order to obtain a frequency dependent "measure" of fluctuations analogous to the notion of standard deviation, we define the "Root Mean Square spectrum" as follows (it has the same units as the signal X(t) and σ_X):

$$S_{XX}^{RMS}(f_i) = 2\sqrt{S_{XX}(f_i)\Delta f}\,. \qquad (3b)$$

The frequencies and dimensionless frequencies are defined as follows:

$$f_i : \text{frequency (Herz) (1 Hz=1 s-1)}; \; T_i = \frac{1}{f_i} : \text{period (seconds)}$$

Let M be either the number of time steps in the data, or the maximum number of lags allowed in the Tuckey filter D(j) (*see next section 3.3.2*). If the Tuckey filter is turned off, one can let M = N-1 where N is the total number of points in the observed signal. Then, we have:

$$f_i = \frac{1}{2M\Delta t_i} : \text{period (seconds)}$$

According to Shannon's sampling theorem (see **Figure** 5), here are the maximum and minimum frequencies that can be "identified" given the time step and the duration of the signal[5]:

$$f_{max} = \frac{1}{2\Delta t} = \frac{1}{T_{min}} \text{ [Hz]} \tag{4a}$$

$$f_{min} = \frac{1}{T_{max}} = \frac{1}{2t_{max}} = \frac{1}{2M\Delta t} \text{ [Hz],} \tag{4b}$$

where $t_{max} = M\Delta t$ (size of the time window). In these relations, Tmin and Tmax are, respectively, the smallest and largest periods that can be represented with this time series ($T_{max} \geq 2T_{min}$). The dimensionless frequency $\hat{f}_i$ is defined by:

$$\hat{f}_i = f_i \times \Delta t \implies \hat{f}_i = \frac{i}{2 \times M \times k_0}; \tag{4c}$$

where ko is the sub-sampling step in time (we always use full sampling with ko = 1 in this work).

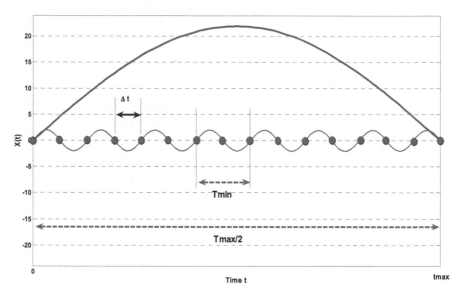

Fig. 5. Schematic illustration of some aspects of signal sampling in discrete time within a limited observation window (which may be due to truncation of a longer time series). Δt : time step (seconds); $t_{max} = M\Delta t$: size of observation window. T_{min} : smallest period that can be identified (Shannon). T_{max} : largest period that can be identified. $f_{max} = 1/T_{min}$ is Shannon's maximum frequency (sampling theorem).

---

[5] Note: Shannon's sampling theorem concerns in fact only the maximum frequency; but nevertheless, there is also a minimum frequency theorem, as shown below.

Statistical Analyses of Pore Pressure Signals in Claystone During Excavation Works at the Mont
Terri Underground Research Laboratory

19

The dimensionless frequency $\hat{f}_i$ belongs to the interval $[0, 1/2]$, that is: $0 \le \hat{f}_i \le 0.5$. Thus, for $k_0 = 1$, we have:

$$\hat{f}_0 = 0 \text{ and } \hat{f}_M = 0.5 \tag{4d}$$

The index "j" in the Fourier spectrum/covariance *Eq. 2* is the label of the *time lag*, defined as follows:

$$j = 0, 1, 2, \ldots, M \le N\text{-}1$$

where the zero time lag (j = 0) is taken into account separately outside the sum in *Eq. 2*. On the other hand, the index "i" is the label of discrete frequency, or of dimensionless frequency ($\hat{f}_i$):

$$i = 0, 1, \ldots, M \le N\text{-}1 \text{ or (see the note just below): } i = 0, 1, \ldots, M, \ldots, N\text{-}1.$$

The number "M" represents the discrete size of the filter truncation window (number of Δt's in the maximum lag considered), while "N" is the total number of data of the observed signal (full time window), and Δt is the constant time step of measurements (possibly after homogenization of time steps). In general, for a filtered spectrum, M $\ll$ N (at most, M=N-1 for the unfiltered spectrum).

Note. Concerning the discrete number of frequencies represented in the filtered spectrum (index "i"), it is possible to impose either a "coarsened" representation of the spectrum with frequencies running from i=1 to i=M (where usually M$\ll$N), or alternatively, to maintain a fine representation of the spectrum with frequencies running from i = 0, 1, ...N-1, regardless of the maximum time lag "M" used in the Tuckey filter "D". In the latter case, one should replace "M" by "N-1" in all of *Eqs. 4* (index "i").

3.3.2 The Tuckey filter ("hanning" filter) for spectral estimation

Finally, note that "s(f)" in *Eq. 2* is an estimator of the spectrum, with a filter function "D". More precisely, "Dj" represents a time-lag filter function; it is applied to the correlation function within the Fourier transform. The weight function D(j) included inside the Fourier transform represents a "smoothing function" or "filter" of Rxx(j) in time lag space (j), where "j" is the discrete index of the dimensional lag time $\tau(j)$. Thus, we can write equivalently D(j) or D($\tau(j)$).

Our particular choice here the "Tukey-2" filter (also called a *"hanning"* filter):

$$D_j = \frac{\left(1 + \cos\left(\dfrac{\pi j}{M}\right)\right)}{2} \tag{5}$$

where the index "j" runs generally over (j = -M,...,-1,0,+1,...,+M), or here over (j = +1,...,+M), in the particular discrete Fourier Transform sum written in *Eq. 2*.

The filter D(j) is necessary in most cases. Indeed, it is necessary to apply smoothing to Rxx(τ) or Rxx(j) because of the sampling errors on the estimated Rxx(j), which is obtained on a finite time window. Thus, it can be shown precisely that the estimation error on the covariance Cxx(j) increases with time lag τ(j): see *Yevjevich (1972, Chap.3)* and *Bartlett (1946)*. For example, in the case of an autoregressive process with exponential covariance function $C_{XX}(\tau) = \sigma_X^2 \exp(-|\tau|/\tau_{COR})$, it has been shown (*Ababou et al. 1988: Appendix 2*) that the variance of sampling error, defined by $\varepsilon_{RMS}^2(\tau) = Var(\hat{C}_{XX}(\tau))/C_{XX}^2(\tau)$, is given by:

$$\varepsilon_{RMS}^2(\tau) = 2\frac{\tau_{COR}}{t_{MAX}}\left(1 - \frac{1}{2}\frac{\tau}{t_{MAX}} - \frac{1}{4}\frac{\tau_{COR}}{t_{MAX}} + \frac{\tau}{\tau_{COR}}\left(1 - \frac{1}{2}\frac{\tau}{t_{MAX}}\right)\right). \tag{6}$$

in terms of continuous lag time (τ). The two parameters involved in this analytical example are the signal's correlation time (τ_{COR}) and the size of the observation window (t_{MAX}). As can be seen, the relative RMS error ε_{RMS} is proportional to $\sqrt{\tau_{COR}/t_{MAX}}$ at zero lag ($\tau = 0$), and ε_{RMS} goes to 100% as $\tau \to t_{MAX}$. This illustrates why it is necessary to apply a filter like D(τ(j)) in order to smooth out the estimated Cxx(τ(j)) or Rxx(τ(j)) at large time lags, and also to smooth out the resulting frequency spectra obtained by Fourier transform of Cxx(τ(j)) or Rxx(τ(j)).

The filter we have chosen here ("Tukey-2", or "*hanning*" filter) adequately smoothes out large time scales (or noise at the scale of low frequency increments in the frequency domain). Experience with hydrological signals indicates that this filter has less variance bias than others (according to *Mangin 1984*, it over-estimates the total variance by no more than 8%). However, other filters may be considered in future: in particular, it should be noted that a "minimum bias" spectral filter has been proposed by *Papoulis & Pillai (2002)*.

Finally, note that the Fourier transform of our time-lag filter D(τ(j)) is itself a "filter" in frequency space (D(f)). Its spectral band should contain a large part of the spectral band of the signal, in order not to waste too much information in this filtering procedure when estimating the spectrum. In our applications, the maximum lag "M" in the D(τ(j)) filter (*Eq. 5*) was adjusted depending on each specific application and objective (time scales and frequencies of interest).

3.3.3 Cross-spectrum, frequency gain, and coherence function

The (dimensional) cross-spectrum of two signals (X(t), Y(t)) is a complex function of frequency which can be described as follows:

$$S_{XY}(f_i) = |S_{XY}(f_i)|e^{\varphi_{XY}(f_i)} \quad \text{or} \quad S_{XY}(f_i) = S_{Re}(f_i) + S_{Im}(f_i) \tag{7a}$$

where

$$|S_{XY}(f_i)| = \sqrt{S_{Re}^2(f_i) + S_{Im}^2(f_i)} \tag{7b}$$

and

Statistical Analyses of Pore Pressure Signals in Claystone During Excavation Works at the Mont
Terri Underground Research Laboratory

21

$$\varphi_{XY}(f_i) = \tan^{-1}\left(\frac{S_{Im}(f_i)}{S_{Re}(f_i)}\right), \tag{7c}$$

are respectively the amplitude (cross)-spectrum and the phase (cross)-spectrum. The latter is defined only in an interval of size π, such as $[-\pi/2, +\pi/2]$. More precisely, however, the phase spectrum can be identified unambiguously within the full interval $[-\pi, +\pi]$ with the equation:

$$\varphi_{XY}(f_i) = sign(S_{Im}(f_i))\cos^{-1}\left(\frac{S_R(f_i)}{|S_{XY}(f_i)|}\right), \tag{7d}$$

as will be demonstrated shortly with a simple test.

Given the complex cross-spectrum Sxy(f), one can also define its real part (or "co-spectrum"), and its imaginary part (or "quadrature spectrum"). These are defined and estimated using the Wiener-Khinchine representation (similar to that used in *Sec. 3.3.1* for the auto-spectrum). The result is expressed below in terms of sine and cosine transforms (where a Tuckey filter "Dj" is inserted):

$$S_{Re}(f_i) = 2\Delta t \sigma_X \sigma_Y \left[R_{XY}(0) + \sum_{j=1}^{M}(R_{XY}(j) + R_{YX}(j))D_j \cos(2\pi j\Delta t f_i) \right] \tag{8a}$$

$$S_{Im}(f_i) = 2\Delta t \sigma_X \sigma_Y \left[\sum_{j=1}^{M}(R_{XY}(j) - R_{YX}(j))D_j \sin(2\pi j\Delta t f_i) \right] \tag{8b}$$

It is useful to define also the dimensional Root Mean Square cross amplitude spectrum, as follows (it is important to note that this RMS cross-spectrum has the units of $\sigma_X \times \sigma_Y$):

$$\left|S_{XY}^{RMS}(f_i)\right| = 2\sqrt{|S_{XY}(f_i)|\Delta f} \tag{9}$$

Finally, we define a spectral coherence function:

$$Coh_{XY}^{RMS}(f_i) = \frac{\left|S_{XY}^{RMS}(f_i)\right|}{\sqrt{S_{XX}^{RMS}(f_i)\,S_{YY}^{RMS}(f_i)}} \tag{10}$$

and a spectral gain, or frequency gain function:

$$G_{XY}^{RMS}(f_i) = \frac{\left|S_{XY}^{RMS}(f_i)\right|}{S_{XX}^{RMS}(f_i)} \tag{11}$$

In the latter equation defining the "gain", it should be noted that the two signals are treated differently: X(t) is considered the input signal, and Y(t) the output signal. Thus, Gxy(f) is the input→output spectral gain. It quantifies, for each frequency "f", the effect G of the signal X(t) on the output signal Y(t). This effect is called the gain.

3.3.4 Synthetic example of cross-spectral analysis on two harmonic signals

Two "synthetic" harmonic signals are used in order to test the auto- and cross-spectral estimates described above (spectrum of each signal, then cross amplitude spectrum, phase spectrum, frequency gain).[6] The two signals, input $X(t)=X_1(t)$ and output $Y(t)=X_2(t)$, are defined as follow:

$$X_1(t) = A_0 + \sum_{k=1}^{4} A_k \cos(\omega_k t + \phi_k) \; ; \; X_2(t) = B_0 + \sum_{k=1}^{4} B_k \cos(\omega_k t + \gamma_k) \qquad (12a)$$

Both signals have four harmonic components, with the same four frequencies $\omega_i = 2\pi/T_i$ (i = 1,2,3,4) with periods T_i = 6 months, 8 days, 24 hours and 12 hours, respectively. In addition, each signal has its own constant mean (A_0 or B_0). The mean may be viewed as the zero frequency harmonic (i = 0). The four amplitudes and phases are different for the two signals. Here is a summary or the values used in this test:

- **Signal $X_1(t)$:** Mean A_0 = 50 [*units of X_1*]; amplitudes (A_i) = (12, 9, 4, 1) [*units of X_1*]; phases (ϕ_i) = ($\pi/4$, 0, $\pi/8$, $\pi/6$) [*rad*].
- **Signal $X_2(t)$:** Mean B_0 = 50 [*units of X_2*]; amplitudes (B_i) = (1, 4, 9, 12) [*units of X_2*]; phases (γ) = (0, 0,0, 0) [*rad*].

In the test presented below, the two synthetic signals are generated during a time span of 1 year (in comparison, the longest period in the signal is 6 months). The time step is Δt = 900 s = 15 mn (in comparison, the shortest period in the signal is 12 hours). Spectral estimations are carried out "unfiltered" (the Tukey filter "Dj" is not applied: we take M = N-1 as the maximum time lag).

The spectral estimation results, shown in *Figure 5* & *Figure 6*, are exactly as expected: they coïncide very closely with the exact theoretical values that we have inferred directly from cross-spectral analysis of these harmonic signals (assuming infinite record $t_{MAX} \to \infty$, and continuous time $\Delta t \to 0$). For reference, here are some theoretical results for each discrete frequency $f(i) = \omega_i / 2\pi$:

$$Coh_{XY}^{RMS}(f_k) = 1; \; \left| S_{XY}^{RMS} \right|(f_k) = \sqrt{A_k} \sqrt{B_k} \; ;$$
$$G_{XY}^{RMS}(f_k) = \sqrt{B_k / A_k} \; ; \; \varphi_{XY}(f_k) = \phi_k - \gamma_k \qquad (12b)$$

For instance, focusing on the frequency of the *24 hour* harmonic, we have the following results (we use *unbiased* covariances, except for phase & gain spectra which are estimated using *biased* covariances):

- the RMS auto-spectrum of $X_1(t)$ should be exactly 4.00 [*units of X_1*], and we find 3.996 from the estimated spectrum (*Fig.7.a*);
- the RMS auto-spectrum of $X_2(t)$ should be exactly 9.00 [*units of X_2*], and we find 8.999 for the estimated spectrum (*Fig.7.b*);

---

[6] Note: this example was developed recently, based on similar tests developed earlier by the authors and collaborators (see acknowledgments).

- the RMS coherence spectrum [*dimensionless*] should be exactly 1 for each of the four harmonics,
  and we do find that result for the estimated coherence spectrum (*Fig.7.c*);
- the RMS amplitude cross-spectrum should be exactly 6.00 [*units of* $\sqrt{X_1}\sqrt{X_2}$], and we find 5.997 for the estimated cross spectrum amplitude (*Fig.7.d*);
- the phase spectrum should be exactly $\pi/8$ [*rad*], and we find 0.391 [*rad*] for the estimated phase (*Fig.6.a*);
- the spectral gain between $X_1(t)$ and $X_2(t)$ should be exactly $\sqrt{9/4} = 3/2$ [*units of* $\sqrt{X_2}/\sqrt{X_1}$], and we find 1.50, the exact result with two significant digits (*Fig.6.b*).

As a final note on spectral estimation, it should be emphasized that the Wiener-Kinchine method used in this work relies on the Fourier transform of auto- and cross-correlation functions, which are themselves estimated in the time domain assuming that the processes $(X(t), Y(t))$ are jointly statistically stationary. Alternatively, however, the power spectrum $S_{XX}(\omega)$ of a signal $X(t)$ could be estimated from the *"periodogram"* $P_{XX}(\omega)$, which is the direct Fourier transform of the signal $X(t_i)$ itself (over a finite time window, in discrete time). Filtering or smoothing will be again necessary, as in the Wiener-Khinchine method. The difference is that the required smoothing can now be performed in frequency space (ω). For instance, the periodogram can be smoothed out using a moving average filter in frequency space, to finally obtain a filtered estimate of $S_{XX}(\omega)$. For more details on this, the reader may consult *Blackman and Tuckey (1958)*, *Yevjevich (1972, Chap.3)*, *Papoulis and Pillai (2002, Chap.12)*, and *Priestley (1981)*.

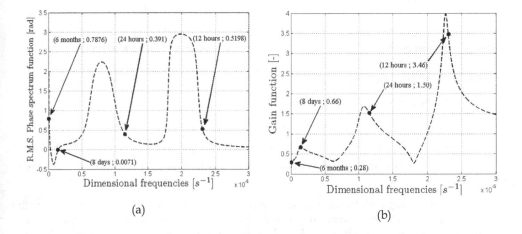

(a)

(b)

Fig. 6. Synthetic test with 4 harmonics: (a) the RMS phase spectrum function [phase difference between $X_1(t)$ and $X_2(t)$, in *radians*]; (b) the RMS spectral gain function [gain between $X_1(t)$ and $X_2(t)$, in *units of* $\sqrt{X_2}/\sqrt{X_1}$].

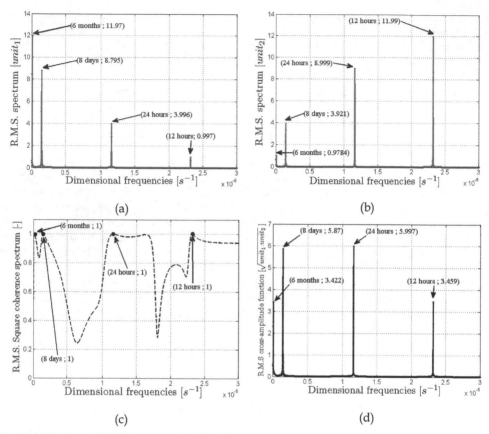

Fig. 7. (a) Estimated RMS spectrum of X_1; (b) Estimated RMS spectrum of X_2; (c) Estimated RMS square coherence spectrum; (d) Estimated RMS cross-amplitude spectrum.

3.4 Multiresolution wavelet analysis

3.4.1 Introduction on wavelets

The analysis of a signal in terms of multi-resolution wavelets uses an orthogonal basis $\{\psi_{j,k}(t)\}$ where "k" represents the temporal translation (position), and "j" represents the dilation or the contraction of time scale, based on a dyadic time discretization of the signal $X(t)$.

The basis being orthogonal, the components of different "scales" are independent from each other. Moreover, each component at a given scale can be traced according to the "time" parameter or "temporal position". This is to be compared with the traditional Fourier analysis, where each Fourier coefficient $C(f)$ represents the overall contents of the signal at frequency « f ». In other words, Fourier coefficients $C(f)$ encapsulate the "f" frequency content of the signal $X(t)$ for all times "t", without representing the possible temporal localization of the frequency content.

On the other hand, the orthogonal wavelet approach makes it possible to clearly highlight non-stationary behaviour and time-localization phenomena (*Labat et al. 2000b*). It also makes it possible to follow the relationship between the various scale levels as a function of time, and to isolate each independent component to study its own evolution. In this sense, these analyses can lead to a better identification of the process responsible for the observed variations. For an extension to wavelet multivariate analyses, or cross-wavelet analyses, see *Labat et al. (2002)*.

We choose here to conduct multi-resolution analysis with discrete orthogonal wavelets, using the wavelet bases of Daubechies. The software employed in this work combines a set of wavelet functions programmed in MATLAB and available on line at the University of Stanford (WAVELAB), and other functions contained in MATLAB's own WAVELET TOOLBOX (some of which are used by Stanford's WAVELAB package).

3.4.2 Wavelet decomposition of a signal on an orthogonal multiresolution basis

In this subsection, we present the decomposition of a signal on a wavelet basis, first for continuous wavelets (non orthogonal bases), and secondly for dyadic multi-resolution wavelets. In the remainder of this text, we will focus essentially on the latter wavelets, for which a compactly supported orthogonal wavelet basis can be constructed (Daubechies basis). The term "dyadic" refers to the fact the scales of multiresolution wavelets vary discretely as powers of "2". For example, if the length of the recorded signal is $t_{MAX} = 64\Delta t$, then there are exactly 7 wavelet scales $2^j\Delta t$ $(j=0,1,...,6)$; the shortest scale is $2^0\Delta t = \Delta t$, the longest scale is $2^6\Delta t = 64\Delta t = t_{MAX}$. The signal can be decomposed in terms of these dyadic time scales. We will define later (*in sub-section 3.4.3*) the concepts of wavelet "detail", "approximation" and "residual", at any given dyadic scale.

3.4.3 Continuous wavelets (a brief introduction)

A signal X(t) is represented formally via wavelet coefficients $C^X_{s,\tau}$ in continuous time as follows:

$$X(t) = \frac{1}{K_\psi} \int_{\sigma=0}^{\sigma=+\infty} \frac{1}{\sigma^2} \left(\int_{\tau=-\infty}^{\tau=+\infty} C^X_{\sigma,\tau}\, \psi_{\sigma,\tau}(t)d\tau \right)d\sigma \qquad (13a)$$

$$\psi_{\sigma,\tau}(t) = \frac{1}{\sqrt{\sigma}}\psi\left(\frac{t-\tau}{\sigma}\right)\ ;\ t\in\mathbb{R};\ \tau\in\mathbb{R};\ \sigma\in\mathbb{R}\text{-}\{0 \qquad (13b)$$

$$C^X_{\sigma,\tau} = \int_{-\infty}^{+\infty} X(t)\psi^*_{\sigma,\tau}(t)dt \equiv \left\langle X(t),\psi^*_{\sigma,\tau}(t)\right\rangle \qquad (13c)$$

where "t" is time, "τ" is a time-shift parameter, and "σ" is a scale parameter (a time scale parameter). It is assumed, for simplicity, that the signal is known for $t\in[-\infty,+\infty]$. *Eq. 13a* gives the decomposition of signal X(t). *Eq. 13b* expresses the (dilated) wavelet function at a given scale σ. *Eq. 13c* gives the wavelet coefficients of the signal: they are the scalar products of X(t) with the (conjugate) dilated and translated wavelet functions $\psi_{\sigma,\tau}{}^*(t)$.

A discrete time analogue of these equations can be constructed as follows: **(a)** replace "t" with $t(i) = i \times \Delta t$ ($i = \ldots, -2, -1, 0, +1, +2, \ldots$) where "i" is the time index; **(b)** replace "τ" with $\tau(k) = k \times \Delta t$ ($k = \ldots, -2, -1, 0, +1, +2, \ldots$) where "k" is the time-shift index; and **(c)** replace "σ" with $\sigma(s) = s \times \Delta t$ ($s = 0, 1, 2, 3 \ldots$) where "s" is the scale index (*discrete but not dyadic -- see further below*). Similarly, a finite time window analogue can also be constructed. Thus, one can deal with the practical case of discrete time step Δt and finite window $[-T_{MAX}/2, +T_{MAX}/2]$ or $[0, T_{MAX}]$.

Different wavelet bases can be obtained depending on the choice of the "mother wavelet" $\psi(t)$. However, in the framework of "continuous" wavelets, it is not possible to find a set of wavelet functions $\psi_{s,k}(t)$ that form an orthogonal basis. For example, the *Morlet* wavelet basis is not orthogonal, implying that there are redundancies in the *Morlet* wavelet decomposition.

3.4.4 Multi-resolution wavelets and dyadic time scales

In contrast, with dyadic multi-resolution wavelets, an orthogonal wavelet basis with compact support can be constructed (*Daubechies 1988, 1991*), leading to multi-resolution wavelet analysis (*below*)[7]. This is made possible by the use of dyadic scales of the form $\sigma(j) = 2^j \times \Delta t$ (powers of 2). Thus, in what follows, let "i" be the time index; "j" the dyadic scale exponent (power); and "k" the time-shift index. The decomposition of X(t) on the multi-resolution wavelet basis takes the form:

$$X(i) = \sum_{j=0}^{j=+\infty} \sum_{k=-\infty}^{k=+\infty} C_{j,k}^X \psi_{j,k}(i) \quad \text{(where "i" can be replaced by "}t_i\text{")} \tag{14a}$$

$$\psi_{j,k}(t_i) = 2^{j/2} \psi(2^j \times t_i - \tau_k) \tag{14b}$$

$$C_{j,k}^X = \sum_{i=-\infty}^{i=+\infty} X(t_i) \psi_{j,k}^*(t_i) \equiv \left\langle X(t_i), \psi_{j,k}^*(t_i) \right\rangle \tag{14c}$$

where:

$j \geq 0$: exponent of dyadic scale dilatation "$2^j \times \bullet$" (or compression if $j \leq 0$);

k: discrete time shift $\leftrightarrow \tau(k) = k \times \Delta t$; i: discrete time $\leftrightarrow t(i) = i \times \Delta t$;

$i \in Z$; $k \in Z$; $j \in N$ (the latter is the discrete dyadic scale)

The $\psi_{j,k}(t)$ basis functions form an orthonormal basis, with their images obtained by time translations and scale dilations/compressions of the mother wavelet $\psi(t)$ [8]. Thus, we have:

---

[7] Two functions f(t) and g(t) are "orthogonal" if their scalar product <f(t),g(t)> is null, where we define <f(t),g(t)> as the integral of f(t)×g(t) over the time domain. A function f(t) has compact support if it is null outside a finite interval [a,b].

[8] In the literature, the decomposition of X(t) is formulated in such a way that only dilatations (or "dilations") are retained, starting from the smallest scale Δt at j = 0, and increasing "j"; but it is possible to formulate this differently in terms of both dilatations and compressions.

$$\int_{-\infty}^{+\infty} \psi_{m,n}(t)\psi_{m',n'}(t)\,dt = \delta_{m,m'}\,\delta_{n,n'} \tag{15}$$

where δ is the classical Kronecker symbol defined by $\delta_{ij}=1$ if i=j and $\delta_{ij}=0$ if i≠j. **Eq. 14a** shows that the discrete time signal X(i) or X(t(i)), of finite energy, can be broken up into a linear combination of translations and dyadic dilatations of the basic functions with adequate coefficients C(j,k) given by **Eq. 14c**. Note that **Eq. 14a** can be viewed as the synthesis of a signal based on wavelet coefficients C(j,k). It can also be interpreted as a succession of approximations of the signal X(i) (in the least-squares sense) by a sequence {Xn(i): n=1,.....N} defined by:

$$X_n(i) = \sum_{j=0}^{j=n-1}\sum_{k=-\infty}^{k=+\infty} C_{j,k}^{X}\psi_{j,k}(i) \ \left(X_n(i) \to X(i) \ \text{ as } \ n \to \infty \right). \tag{16}$$

The above equations constitute the conceptual basis of multi-resolution wavelet analysis. For more details on the theory, see *Mallat (1989)*. Below, we focus on some useful consequences of the orthonormal wavelet decomposition above, introducing the wavelet approximation + residual, and interpreting the wavelet "approximation" as a filter of given dyadic scale.

3.4.5 Multiresolution wavelet decomposition: approximation A(t), residual R(t), and detail D(t) or component C(t)

In this subsection, we define the important concept of wavelet "detail" or "component", i.e., and we show how to extract from the signal X(t) a wavelet "component" with any given dyadic scale. It is important to note that only dyadic scales can be extracted; however, this is the price to pay for the other "good" properties (compact support and orthogonality of the decomposition). We also present the decomposition of signal X(t) into an "approximation" A(t) plus a "residual" R(t), both defined at any given cut-off scale (again, the cut-off scale can only be dyadic). Remarkably, the wavelet "approximation" can be viewed as a "filtered" version of the original signal: it contains only longer time scale components (longer than the given dyadic cut-off scale).

3.4.6 Orthogonal decomposition of X(t) - wavelet components, approximation, residual

Let us assume for simplicity that the (finite) discrete signal $X(t_i)$ has a dyadic length ($N=2^M$), the number of time steps is a power of 2 (dyadic). The multi-resolution wavelet decomposition above can be expressed as a succession of nested *"approximations"* A_m, corresponding to increasing *scales* *"m"* (thus, "m" can designate a chosen dyadic time scale cut-off).

The difference between the signal $X(t_i)$ and its approximation of order *"m"* is the *"residual"*. The difference between the approximations of order (m) and (m+1) is called the *"detail"* of order (m+1).

Intuitively, the *approximation* corresponds to the smoothed image of the signal, while the *"detail"* is a particular *wavelet component* that highlights the irregularities of the signal at a certain scale.

Starting from *Eq. 14a*, we obtain at each selected dyadic scale "m" the "wavelet detail" $D_m(t_i)$:

$$D_m^X(t_i) = \sum_{k=-\infty}^{k=+\infty} C_{m,k}^X \Psi_{m,k}(t_i) \iff D_X^m(i) = \sum_{k=-\infty}^{+\infty} \langle X, \Psi_{m,k} \rangle \Psi_{m,k}(i) \tag{17a}$$

where "t(i)" can also be replaced by "i". The approximation A_m of the signal $X(t_i)$ at dyadic scale resolution "m" (cut-off scale "m"), and the residual R_m at the same scale "m", are then given by:

$$R_m^X(t_i) = \sum_{j \le m-1} D_j^X(t_i); \ A_m^X(t_i) = X(t_i) - R_m^X(t_i) \Rightarrow A_m^X(t_i) = \sum_{j \ge m} D_j^X(t_i) \tag{17b}$$

Thus, at any dyadic scale "m", the signal is the sum of its approximation and residual:

$$X(t_i) = A_m^X(t_i) + R_m^X(t_i) \tag{17c}$$

Using the latter equations to calculate the approximation of order (m-1) leads to:

$$A_{m-1}^X(t_i) = A_m^X(t_i) + D_{m-1}^X(t_i) \iff A_{m+1}^X(t_i) = A_m^X(t_i) - D_m^X(t_i) \tag{17d}$$

As "m" increases, the A_m's become coarser and coarser approximations. The above equation shows that the "coarser" approximation A_{m+1} is obtained from A_m by subtracting the detail D_m. Note that the zero order residual R_0 is null by construction. The zero order approximation A_0 is identical with the original signal. Thus, we have, with our index conventions:

$$R_0^X(t_i) = 0; \ R_1^X(t_i) = D_0^X(t_i); \ ...; \ R_M^X(t_i) = X(t_i) - D_M^X(t_i); \ R_{M+1}^X(t_i) = X(t_i) \tag{17e}$$

$$A_0^X(t_i) = X(t_i); \ A_1^X(t_i) = X(t_i) - D_0^X(t_i); ...; A_M^X(t_i) = D_M^X(t_i) \tag{17f}$$

where M is the exponent of the largest dyadic scale. For example, M = 6 if the signal length N is in the range $64 \le N \le 127$. The dyadic wavelet decomposition is performed without loss of information if N is a power of "2", such as N = 64; but if N = 127 the loss is maximal.

3.4.7 Theoretical interpretation of multiresolution wavelets (nested spaces and filters)

For completeness, note that a scale function $\varphi(t)$ is also introduced. It can be shown that the "approximation" can be directly decomposed orthogonally over the set of dilated/translated scale functions $\varphi_{j,k}(t)$ (*not shown here*). The scale function $\varphi(t)$ is also called the *father wavelet*, to be distinguished from the *mother wavelet* $\psi(t)$ used above in the decomposition of *Eq. 17a*.

Theoretically, approximations A_m and details D_m are each decomposed into bases of φ's and ψ's, respectively, which form nested spaces called V_m and W_m, respectively. Considering a given scale "m", and recalling that $m \ge 1$ ("m" increases for increasing dyadic scale $2^m \times \Delta t$), it can be seen that the "detail" D_m represents the wavelet component of scale "(m)", the residual R_m incorporates the sum of all scales strictly smaller than "m", and the approximation A_m incorporates all scales greater or equal to "m" (*note: the approximation "m" includes the scale "m" itself*).

Statistical Analyses of Pore Pressure Signals in Claystone During Excavation Works at the Mont
Terri Underground Research Laboratory

29

The nested spaces V and W are related by $V_{m+1} = V_m$-W_m, as indicated by *Eq. 17d*. In fact, *Eq. 17d* describes a sequence of multi-resolution filters (the A_m's) and their residuals (the R_m's).

3.4.8 Remarks on multi-resolution approximation as a filter

The process of selecting a given wavelet scale "m" is similar to passing a "band pass" filter on the signal. The wavelet "approximation" A_m can be viewed as a "low pass filter" of given dyadic scale. This filter is somewhat similar to a moving average filtering of the original signal. It was shown empirically in some cases that the two filters are quite close. See for instance *Wang et al. (2010)* concerning water level signals in coastal & beach hydrodynamics. The advantage of the wavelet approximation as a "filter" is that it has the property of being orthogonal to its residual.

3.4.9 Final remarks on wavelets notations and terminology (pitfalls)

The presentation above is certainly too brief to be complete. Therefore, let us emphasize a few more points in order to avoid any possible confusion with multiresolution wavelets.

i. The wavelet "detail" $D_m(t_i)$ defined above can also be seen as a wavelet "component", to be denoted $C_m(t_i)$: this notation designates the wavelet component (C) at the mth-dyadic scale (m). The sum of all components (over all scales "m") yields the signal itself. *Note*: from now on, we will prefer to use this notation, component $C_m(t_i)$, rather than $D_m(t_i)$.

ii. For each given scale "m", the wavelet "component" $C_m(t_i)$ is a discrete time signal. This signal (component C(t)) *should not be confused* with the "wavelet coefficients C_{jk}", defined earlier, where "k" designates a time shift parameter. The "content" of signal $X(t_i)$, at dyadic scale "j", is the sum over all time shifts "k" of the product $C_{jk} \times \psi_{jk}(ti)$…The coefficient Cjk alone is not a time signal.

iii. Multi-resolution wavelets can be presented differently in other texts, with other index conventions and notations. For example, the dyadic time scale labeled "m" can either increase or decrease with index "m" *(here we have chosen that the scale increases with "m")*. Also, some authors choose *(1/scale)* instead of *(scale)* in their visualisations of the wavelet decomposition in scale-time space.

3.4.10 Example application: Multiresolution wavelet analysis of a synthetic signal

Given the limitations of Fourier analysis, multiresolution analysis is particularly useful as a tool for analysis, synthesis, and/or compression, of complex *non stationary* signals. For this reason, we present briefly a synthetic example involving a signal that has a simple harmonic component added to a strongly non stationary mean or trend, as follows:

$$X(t_i) = \begin{vmatrix} 1+0.5\sin(2\pi t_i/8) & \text{if } t_i < 15 \\ 1-(t-16)/8+0.5\sin(2\pi t_i/8) & \text{if } t_i \geq 15 \end{vmatrix} ; \ t_i = i\Delta t ; \ i = (0,1,2,...,31) ; \ \Delta t = 1 \quad (18)$$

This signal has an exactly dyadic length $N = 32 = 2^5$ (the dyadic exponent is M = 5).

Figure 8 (a,b,c,d) shows the signal X(t) and three different decompositions using the wavelet bases of Haar (Daubechies 2), Daubechies 4, and Daubechies 20. In each case, the six wavelet components $C_i(t)$ are shown (i = 0, 1, ..., 5). It can be seen that component $C_2(t)$ tends to capture the sinusoïd of period T=8. The corresponding dyadic scale is $S_j = 2^j \Delta t = 4$ with j=2. Therefore, we see that the wavelet scale corresponds in this case to the half-period of the signal. On the other hand, the non stationary mean level is captured by component $C_4(t)$, with scale $S_j = 2^j \Delta t = 16$, which corresponds to the half-duration of the non stationary signal. Note also that the components become smoother as the degree of the wavelet basis increases. The Haar wavelet is well suited to capture abrupt changes of mean level, but not for capturing smoother type of non stationarity.

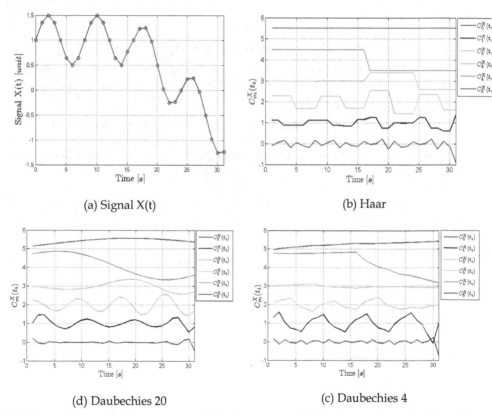

(a) Signal X(t)   (b) Haar

(d) Daubechies 20   (c) Daubechies 4

Note: in graphs (**b, c, d**), each component $C_i(t)$ is shown on a shifted ordinate axis positioned at y = i (i=0,1,2...); this allows for a better visualization of the different components on a single graph; apart from this shift, the ordinate scales are not distorted (same for all components $C_i(t)$).

Fig. 8. The signal X(t) and its dyadic components $C_i(t)$ extracted from three different multiresolution bases (Haar, Daubechies 4 and Daubechies 20). (**a**): The synthetic non stationary signal X(t) to be analysed. (**b**): The decomposition of X(t) into components $C_i(t)$ using Haar (Daubechies 2) mother wavelet. (**c**): The decomposition of X(t) into components $C_i(t)$ using Daubechies 4 mother wavelet. (**d**): The decomposition of X(t) into components $C_i(t)$ using Daubechies 20 mother wavelet.

Statistical Analyses of Pore Pressure Signals in Claystone During Excavation Works at the Mont
Terri Underground Research Laboratory

31

Figure 9 (a,b) shows a construction of the successive approximations and/or residuals of
signal X(t) for the lowest degree wavelet (the Haar wavelet).

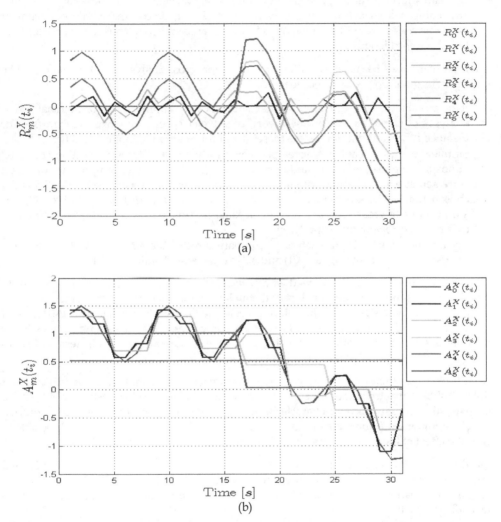

Fig. 9. Plots of successive residuals and approximations using the Haar mother wavelet
(lowest degree wavelet). **(a):** Plot of successive residuals $R_0(t) = 0$, $R_1(t) = C_1(t)$, ..., and
$R_5(t) = X(t)-m_X = X(t)-A_5(t)$ (note that $R6(t) = X(t)$, not shown). **(b):** Plot of successive
approximations $A_0(t) = X(t)$, $A_1(t) = X(t)-C_1(t)$, ..., and $A_5(t) = X(t)$..

3.4.11 Methods and techniques of wavelet analyses and cross-analyses

The above multi-resolution wavelet theory can be used to study different aspects of the
signals, depending on the objectives:

- One can study the *"detail"* or (perhaps more interestingly) the *"residual"* of a signal X(t) at any selected dyadic scale "m". The analysis can be performed for several signals X(t), Y(t). One can compare for instance the details of air and pore pressure at scale 24h.
- One can select dyadic components from one or several signals, and these can be studied and compared as a function of dyadic scale ($2^m*\Delta t$) across the whole spectrum of available scales, from finest scale Δt to largest dyadic scale (which is never smaller than half the record length).

Wavelets can also be used in different ways for cross-analyses of two signals like atmospheric and pore pressure. For instance:

- It is possible to develop a somewhat sophisticated approach for orthogonal cross-analyses of signals based on the theory of wavelet spectra and cross-spectra (see for instance *Labat 2005* and references therein), but this theory will not be used here.
- Simpler analyses can be developed by studying the cross-correlation between atmospheric pressure (Patm) and pore pressure (P). The wavelet component of interest is treated as a signal, and cross-component correlation analysis can be conducted at the chosen time scale: see *Labat et al. (1999 a,b); Wang et al. (2010); Möri et al. (2011);* for applications of cross-wavelet analyses in karst hydrology, beach hydrodynamics, and clay rock hydrogeology, respectively.
- Yet another practical approach is to compare directly the amplitudes of the wavelet components of the two signals X(t) and Y(t) at the same dyadic scale "m".

The latter alternative may be viewed as a *"simplified cross-wavelet approach"*. It is a way to combine two "simple wavelet analyses" to finally obtain a simplified but practical cross-analysis of the two signals at the selected scale. In this work, we will use this type of approach *to identify semi-diurnal earth tide effects* in pore pressure data, leading to an estimate of the elastic "Ss" as the ratio of tidal strain / pore pressure amplitudes in terms of semi-diurnal wavelet components.[9]

In *Sections 4 & 5*, this simplified cross-wavelet approach will be applied to obtain a constant global "Ss" and also to quantify the effect of excavation on a more local estimate of Ss(t). To achieve this, we will need to characterize the time varying amplitude of a semi-diurnal wavelet component (time scale of moon induced earth tides). In his way, we will be able to "see" the effect of the moving excavation front on the semi-diurnal wavelet "signal".

For this task, we need a new tool that can evaluate statistically the "time-modulated amplitude" of a signal. This tool is available already (although we apply it in an unusual way, in combination with wavelets): it is the theory of *"statistical envelopes"*, presented in the next sub-section below.

3.5 Statistical envelope analysis of a signal ("modulated amplitude")

The *Cramer-Leadbetter* theory[10] and the related *Hilbert Transform*[11], are the theoretical tools that enable us to define, mathematically and statistically, the "modulated amplitude" (or the

---

[9] In fact, our current implementation of this general idea to the Mont Terri site is further simplified, using constant tidal amplitude because of the lack of direct earth dilatation record at the site.

[10] Cramer-Leadbetter theory in a nutshell: see *Box N°1*.

Statistical Analyses of Pore Pressure Signals in Claystone During Excavation Works at the Mont
Terri Underground Research Laboratory

33

"envelope") of a fluctuating and evolutionary pressure process. The theory of statistical envelopes is well established for zero mean stationary random processes (*Vanmarcke 1983; Veneziano 1979*). This theory, together with the associated Hilbert Transform, is particularly useful for characterizing relatively narrow-band signals, e.g. ocean & coastal waves (*Huang et al. 1999*).

But the present study (see **Section 5**) suggests that the "statistical envelope" can also be useful in a variety of other cases in geophysics, provided it is combined with carefully selected "components" of the signals (e.g., here, the semi-diurnal wavelet component of pore pressure). In fact, the mathematical theory of statistical envelopes is established only for stationary processes, but as will be seen in **Section 5**, using wavelets, we will get rid of the strong non stationary mean trend due to excavation. This is naturally accomplished by using multi-resolution wavelets at specific dyadic scales. The dominant scale of the mean trend is ~11days, while the wavelet scale selected for analysis is semi-diurnal (scale ≈ 12hours). The effects of the 11 day trend due to excavation are still "visible" as a time varying "modulated" amplitude in the semi-diurnal wavelet component $C_{12HOURS}(t)$, which we capture using envelope theory. Thus, our use of the theory of envelopes and Hilbert Transform in the present context, with semi-diurnal pressure fluctuations superimposed on a strong nonlinear trend due to excavation, seems rather new in the literature.

Cramer-Leadbetter theory in a nutshell.

For a zero mean (detrended) stationary random process X(t), one defines first a dephased "sister" process S(t). Based on X(t) and S(t), one obtains a symmetric double envelope ±R(t) such that:

$R(t)=(X^2(t)+S^2(t))^{1/2}$.

The envelope is usually much smoother than the original process X(t).

In the special case where X(t)=sin(ωt), we obtain S(t)=cos(ωt), and then R(t) = sin²(ωt)+cos²(ωt) = 1, as one may expect. Indeed, it should be obvious that the envelope of the sinusoïdal signal X(t)=sin(ωt) is just a pair of horizontal lines of ordinates -1 and +1.

Hilbert transform in a nutshell.

The Hilbert transform is an integral transform of the form:

$$\hat{X}(t) = H(X(t)) = (h \otimes X)(t) = \int_{-\infty}^{+\infty} X(t)h(t-\tau)d\tau = \frac{1}{\pi}\int_{-\infty}^{+\infty} \frac{X(t)}{t-\tau}d\tau .$$

For a real-valued signal X(t), H(X(t) is essentially the convolution of the signal X(t) by the kernel, or transfer function, h(t) = 1/(πt). Note: X(t) = sin(t) ⇒ H(X(t)) = -cos(t).

The Hilbert transform can be implemented in a complex framework as follows (as in the Matlab software): let X(t) the real valued process; let Y(t) = H(X(t) its real Hilbert transform. Then the complex Hilbert transform of signal X(t) is defined as Z(t) = X(t)+i.Y(t) where **i** is the pure imaginary number (**i** = √-1). The above defined envelope of X(t) is then obtained as:

$$R(t) = \pm\sqrt{X(t)^2 + H(X(t))^2} = \pm\sqrt{X(t)^2 + Y(t)^2} = \pm|Z(t)| .$$

The Hilbert transform is implemented numerically for discrete time signals in MATLAB.

Box 1. Envelope theory in a nutshell

---

[11] Hilbert transform in a nutshell see *Box N°1*.

Rather than present here the full theory of statistical envelopes, we refer the reader to the *above-cited references* for the theory, and to the *summary* presented here in Box N°1. The reader who is not directly interested in the theory can move on to **Section 5** for application of statistical envelope + wavelet analysis, to the non stationary evolution of semi-diurnal fluctuations of (relative) pore pressure during the excavation of Mont Terri gallery Ga98 (*"mine by test"*).

As mentioned earlier, the final objective and result of this analysis is to estimate the evolution of specific storativity "Ss(t)" in the EdZ of the clay rock. This is done by using the tidal dilatation model of Bredehoeft relating tidal strain / relative pore pressure. The originality of this "envelope" approach should be emphasized. The amplitude of the semi-diurnal wavelet component is estimated as a modulated envelope, which varies more smoothly than the signal itself. The resulting storativity coefficient is therefore a time-varying coefficient, Ss(t), which reflects the mechanical disturbance due to the evolving excavation front, over short time scales (days, weeks).

3.6 Physically-based hydro-mechanical models to characterize the clay rock from pressure signals (earth tides and barometric effect)

The statistical methods of signal analysis are complemented or combined with physically based, but *"simplified" hydro-mechanical models.* These are used for hydrogeologic interpretation of the pressure signals (pore pressure and atmospheric pressure), and they are also related to inferred earth tide strain fluctuations, with the aim of characterizing quantitatively some of the hydrogeologic properties of the porous clay rock (Ss, Φ).

Specifically, the simplified hydro-mechanical models used here relay on assuming both water and the prous matrix (bulk) to be elastically compressibility, and on accepting Terzaghi's 1936 effective stress concept. The corresponding models can be traced back to *Jacob (1940)* for the barometric efficiency model; and to *Bredehoeft (1967)* for the earth tides model (also *Jacob 1940* and *Cooper 1966*). These physically-based models both assume that the total stress can be decomposed additively into effective stress between grains, and pressure (fluid stress), according to the Terzaghi concept. This implies that the grains of the claystone are "incompressible". The more general Biot poroelasticity theory can account for the case where individual grains are compressible (see for instance *Hsieh et al. 1988* concerning specific storativity).

In this work we use Terzaghi based hydro-mechanical models. They are outlined below in *Box N°2*, where the final results are expressed in terms of:

- The specific (elastic) storativity Ss [m⁻¹];
- The effective (elastic or dynamic) porosity Φ [m³/m³].

4. Long time scales: statistical structure of pressure signals and characterization of time averaged properties (specific storativity Ss; elastic porosity φ)

This section is devoted to the analysis of long records (months to years) of observed air pressure and relative pore pressure signals, for the purpose of characterizing global and average hydro-geologic properties of the clay stone (global in space, and also, average in time)…Statistical analyses based on correlations, spectra, and wavelets, will enable us to detect and characterize in particular the diurnal relation between relative pore pressure and air pressure (leading to a "barometric efficiency"), and the half-diurnal effect of earth tides (leading to a characterization of compressibility and/or specific storativity).

The purpose of this "box" is to summarize the simplified, but physically-based, hydro-mechanical models. These models are used in combination with statistical analyses in order to interpret pressure signals in terms of rock properties (elastic storativity Ss, effective elastic porosity Φ).

Tidal strain - astronomic and global earth (the semi-diurnal component)

Astronomic movements of Earth, Moon, Sun, induce fluctuations of the gravimetric field g(t), which has an impact on deformations of the earth crust (displacements, strains). Focusing on the M2 harmonic (Moon semi-diurnal), the volumetric strain fluctuations take the form:

$\varepsilon(t) = \Delta\varepsilon.\sin(\omega_{M2}t + \phi)$ where $\Delta\varepsilon$ is the tidal strain amplitude [m³/m³].

Concerning M2, it is the principal lunar earth tide, and its period is more precisely 12h 25 mn. *Takeuchi (1950)* gives: $\Delta\varepsilon = 0.5 \times W_2 / (ag)$, where W2 is the tidal potential of M2, "a" is the radius of the earth, and "g" is gravity. One may take $W2/g \approx 10$ for M2. The constant 0.5 was computed by Takeuchi from earth crust mechanics data (Lamé stiffness moduli, Love number). At the scale of a geologic unit (as opposed to the global scale of the earth), *Melchior (1978)* estimates $\Delta\varepsilon \approx 2 \times 10^{-8}$ $[m^3 / m^3]$ (value adopted *in this work* in the absence of direct measurements).

Tidal strain and elastic specific storativity Ss of a geologic unit - the *Bredehoeft* model

Specific storativity Ss : Ss is in units of [m⁻¹] $\leftrightarrow$ Ss/(ρg) is in units of [(m3/m3)/Pa].

$S_S[m^{-1}] = (\delta V_{MP}/V_{MP})/\delta H$ where V_MP is the volume of deformable bulk Porous Matrix,

and δH is total head variation, with head H defined by: $H = \dfrac{p - p_{ATM}}{\rho_w g} + z$. Equivalently:

$S_S/(\rho g)[Pa^{-1}] = (\delta V_{MP}/V_{MP})/\delta p_{REL}(t)$, with $p_{REL}(t) = p(t) - p_{ATM}(t)$ (relative pressure).

Using Terzaghi's 1936 effective stress concept, assuming constant total stress, and assuming elastic compressibility of water (β=1/Kw) and of the bulk porous matrix (α=1/Kb...) leads to:

$S_S = \rho_w g (\alpha + \Phi\beta) = \rho_w g \left(\dfrac{1}{K_B} + \dfrac{\Phi}{K_W} \right)$ (where Kw $\square$ 2 GigaPa for water).

The *Bredehoeft (1967)* model relates tidal strain to relative pressure fluctuations (or $\square$H) via Ss:

$\delta\varepsilon_{TIDAL} \approx S_S \delta H$, or equivalently: $\delta\varepsilon_{TIDAL} \approx (S_S/(\rho g)) \rho g \delta H = (S_S/(\rho g)) \delta p_{REL}$.

Barometric efficiency B - classical model (*compressibility & effective stress*)

Without going into details (see *Freeze and Cherry 1979*), we present the result obtained with the effective stress hypothesis (*Terzaghi 1936*), in the case where the total stress is held constant, and assuming here also that our hydro-geologic unit behaves like a (*partially*) confined aquifer:

$\delta p_{REL}(t) = -B \times \delta p_{ATM}(t)$, with the barometric efficiency B given by: $B = \beta\Phi/(\alpha + \beta\Phi)$.

Using the previous expression for Ss in terms of (α,β) compressibilities $\Rightarrow B = \beta\Phi/(S_s/(\rho g))$.

In this work, barometric efficiency is re-interpreted statistically as the spectral gain at frequency 1/24h. The above equation shows that barometric efficiency B, specific elastic storativity Ss, and effective elastic porosity Φ are interrelated (the compressibility β=1/Kw of water is known).

Box 2. Summary of physically based hydro-mechanical models

Long records with many time steps (tens of thousands or much more) usually have non stationary structures at various scales. In this section, the undesired non stationary structures are partially eliminated by the selection of specific peak frequencies (in Fourier spectra), and/or by the selection of pertinent dyadic "scales" via multi-resolution wavelet analysis. We seek here only global characteristics of the signals (global amplitudes, etc.).

As explained earlier in the methodology *Section 3*, in order to achieve these goals (hydrogeologic characterization), we use simplified hydro-mechanical "models" to account for the porosity "Φ", the compressibility of water, and the compressibility of the bulk porous matrix. This leads also to the related concept of elastic specific storativity "Ss", based on Terzaghi's decomposition of total stress into effective stress+pressure. The "Ss" coefficient represents an equivalent elastic coefficient for the water filled poro-elastic rock. See the "outlook" comments in *Section 6* for a discussion on the limitations and possible extensions of these hydro-mechanical models.

4.1 A first example: statistical analyses and interpretations of "clean" records of raw (un-processed) pressure signals over~1 month

The above-mentioned statistical techniques are first illustrated here using only relatively short sections of the pore pressure signal PP1(t) at Mont Terri, with constant time step and no data gaps (no reconstitution required): see *Figures 10, 11, 12, 13*.

The unprocessed pore pressure signal is analysed in terms of:

1. its frequency spectrum (shown in *Figure 11*), and
2. its wavelet component at nearly semi-diurnal time scale (shown in *Figure 13*).

4.1.1 Specific storativity from earth tides (using semi-diurnal wavelets)

We estimate in this subsection the month-scale specific storativity "Ss" from semi-diurnal wavelet component of relative pore pressure (strain/pressure -- earth tide effect).

The specific storage coefficient quantifies the volume of water stored per unit volume of porous domain and per unit variation of hydraulic head. It can be estimated from earth tide amplitude using the relation of Bredehoeft (1967), based on a simplified model that takes into account bulk deformation effects due to earth tides in the poro-elastic water filled porous medium:

$$S_s = \frac{|\Delta \varepsilon|}{|\Delta h|} \tag{19}$$

where S_S is specific storativity (m^{-1}), $|\Delta \varepsilon|$ is the amplitude of volumetric strain fluctuations related to the M2 semi-diurnal earth tide, and $|\Delta h|$ is the amplitude of relative pressure head fluctuations (estimated from the semi-diurnal wavelet component of relative pressure). The strain amplitude $|\Delta \varepsilon|$ was estimated by *Takeuchi (1950)* based on astronomic earth tides and earth elasticity, then later on (in the 1960's) by Melchior and others for hydro-geologic formations: see *Melchior (1978)* (originally in *1960), Bredhoeft (1967),* and others: $|\Delta \varepsilon| = 2 \times 10^{-8} \ m^3/m^3$.

Statistical Analyses of Pore Pressure Signals in Claystone During Excavation Works at the Mont
Terri Underground Research Laboratory

37

Note that the site specific signal of volumetric strain $\Delta\varepsilon(t)$ is not available, so its amplitude is only estimated indirectly (from data surveys in the literature). However, our method can be extended to the case where $\Delta\varepsilon(t)$ is available, by applying a semi-diurnal wavelet analysis on the $\Delta\varepsilon(t)$ signal.

The amplitude of relative pressure head $\Delta h(t)$ is obtained from multi-resolution wavelet analysis of $P(t)-P_{ATM}(t)$. The wavelet component with dyadic time scale close to semi-diurnal is isolated, and its amplitude is evaluated using either of the following "norms":

- the RMS (Root-Mean-Square) standard deviation: "*norm 2*" or $\|\ \|_2$; or else

- the absolute mean deviation ("*norm 1*" or $\|\ \|_1$).

Note:

$$\sigma^{(p)} = \left(\frac{1}{N} \sum \left| X(i) - \overline{X} \right|^p \right)^{1/p} \Rightarrow \sigma^{(1)} = \frac{1}{N} \sum \left| X(i) - \overline{X} \right| \text{ and } \sigma^{(2)} = \sqrt{\frac{1}{N} \sum \left(X(i) - \overline{X} \right)^2} \qquad (20)$$

Figure 12 shows the evolution of the components according to time and scale. The goal is to isolate the component corresponding to earth tides (the semi-diurnal component): as indicated on this figure, we have chosen the component of dyadic scale $T = 8h$ (close to 12h). Note: in other calculations, we have attempted to improve the accuracy of semi-diurnal scale selection by computing the mean of the dyadic 8h and 16h wavelet components: $(C_{8H}(t)+C_{16H}(t))/2$.

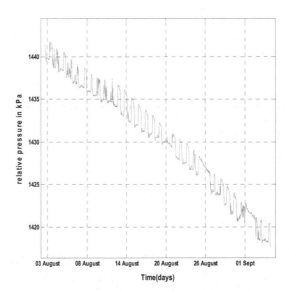

Fig. 10. Signal of the raw relative pore pressure PP1(t) (kPa) at Mont Terri (from 02/08/2002 to 04/09/2002). Time span: 15 month. Time step of raw pressure signal: $\Delta t = 30$ min (and k=1).

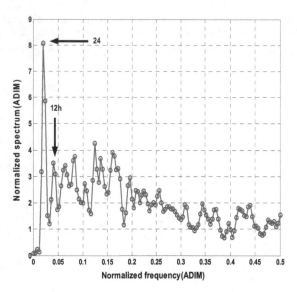

Fig. 11. Reduced spectrum of ΔPP1(t), the first difference of relative pore pressure (kPa), at Mont Terri. Time span: ~1 month (from 02/08/2002 to 04/09/2002). Time step of raw pressure signal: Δt = 30 min (and k=1).

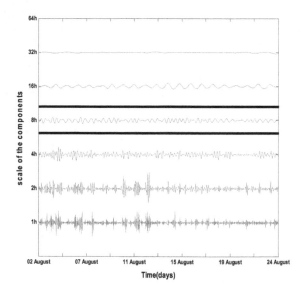

Fig. 12. Multiresolution wavelet analysis : several dyadic time scale components are shown for the raw relative pore pressure PP1(t) at Mont Terri. Duration: 02/08/2002 to 22/08/2002. Time step: Δt = 30 min (and sampling step: k=1).

Statistical Analyses of Pore Pressure Signals in Claystone During Excavation Works at the Mont
Terri Underground Research Laboratory

39

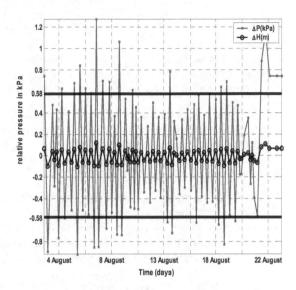

Fig. 13. Multiresolution wavelet analysis : dyadic time scale component 8h (near 12h) for the raw signal of relative pore pressure PP1(t) at Mont Terri. Duration: 02/08/2002 to 22/08/2002. Time step: Δt = 30 min (k=1). Amplitude of fluctuations: Δh = 0.058 m.

Figure 13 shows the evolution of the selected wavelet component (in this case, $C_{8H}(t)$). We have determined numerically the amplitude of the fluctuations of this wavelet component, e.g., its absolute mean deviation. The result can be expressed in terms of relative pressure head: for instance it was found that the absolute mean amplitude ("norm1") of semi-diurnal pressure head is Δh = 0.058 m. We can then calculate specific storativity Ss by applying the previous relation (*Eq. 19*). The results are the following, using either "norm1" or "norm2" for Δh:

Δh(norm 1)=**0.058 m**; S_s(norm 1)= **3.4×10⁻⁷ m⁻¹**
Δh(norm 2)=**0.064 m** ; S_s(norm 2)=**3.1×10⁻⁷ m⁻¹**

4.1.2 Effective elastic porosity Φ from barometric efficiency (using spectral gain)

We now estimate month-scale effective porosity from the diurnal frequency gain between relative pore pressure and air pressure (barometric efficiency).

The diurnal (and other) components would be due in part to barometric pressure effects. The effect of barometric pressure fluctuations on observation wells in closed, confined aquifers, is well known. The piezometric head in the aquifer is related linearly to atmospheric pressure (assuming linear poro-elastic behavior) via a coefficient named "barometric efficiency" (B or B_E).

Although the geometry, boundary conditions, and hydraulic configuration of the Mont Terri claystone appear quite different from those of a confined aquifer, we nevertheless propose to retain the same linear relationship. This makes it possible to calculate effective porosity, according to the relations described in *Section 3.6* and Box N°2 (the following is *Jacob's 1940* relation):

$$\Phi = \frac{E_w \times B}{\rho \times g} \times S_S \qquad\qquad (21)$$

where Φ or Φ_{EFF} is the effective elastic porosity (m³/m³), B the barometric efficiency which expresses the elastic response of the system, ρ the density of water, g the acceleration of gravity, E_W (or K_W) is the stiffness modulus of water ($E_W = K_W \approx 2.05$ GigaPascals).

In order to calculate barometric efficiency B, a statistical cross-spectral analysis of the relation between atmospheric pressure and relative pressure was developed, with 30 minute time steps, to determine the reality of the effect of barometric pressure. We carried out the cross-spectral analyses between barometric pressure and relative pressure in borehole BPP-1 (from the PP niche, which contains two rooms, PP1 and PP2).

The estimated spectral gain function (in frequency space) was finally used to determine barometric efficiency (B) by looking at the value of the spectral gain at diurnal frequency, as shown in *Figure 14*. From this spectral gain, barometric efficiency is determined, and porosity is calculated with *Eq. 21* just above, using the previous estimate of specific storativity (Ss) from semi-diurnal wavelet analysis. We obtain in this fashion the following results:

$$\phi_{EFFECTIVE}\ (\text{norm } 1) \approx 6\%;\ \phi_{EFFECTIVE}\ (\text{norm } 2) \approx 5.4\%$$

Recall that this procedure to obtain "Φ" (by spectral analysis of barometric gain at diurnal frequency) is combined with the previous procedure to estimate "Ss" (by wavelet analysis of semi-diurnal earth tide effects). In fact, the two procedures do not need to be implemented sequentially. They can be viewed as a system of two equations (*Eq. 19* and *Eq. 21*) with two unknowns (Ss, Φ). The two unknowns could have been determined both at once (alternative approach).

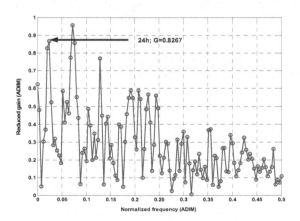

Fig. 14. Reduced spectral gain *vs.* normalized frequency f×Δt (the maximum value of f×Δt is 0.5, in agreement with Shannon's sampling theorem). This spectral gain function was obtained by cross-spectral analysis of the raw signals of atmospheric pressure Patm(t) [kPa] and relative pore pressure PP1(t)-Patm(t) [kPa]. Signal duration: about 1 month (from 02/08/2002 to 04/09/2002). Time step: 30 mn. The gain at diurnal frequency (1/24h) is indicated by the arrow; it is used to estimate barometric efficiency.

Statistical Analyses of Pore Pressure Signals in Claystone During Excavation Works at the Mont
Terri Underground Research Laboratory

41

4.2 Long term analysis of pressure signals over more than 1 year (after pre-processing)

The same techniques illustrated in the previous *Section 4.1* are now being applied to a much longer - but *pre-processed* - joint record of absolute pore pressure and atmospheric pressure (p(t), p_{ATM}(t)). The time span of the processed pair of signals is about 15 months *(from 29 January 2004 to 12 April 2005)*.

Note that, initially, only shorter pieces of "clean" signals were available during that time window. Note also that the pre-processing procedures (reconstructing, homogenizing, etc.) must be implemented jointly on p(t) and on p_{ATM}(t) over the same time window, since the relative pressure signal must be obtained from p_{REL}(t) = p(t)-p_{ATM}(t), and cross-analysis of (p_{REL}(t), p_{ATM}(t)) must be implemented to identify the barometric effect. A flow-chart depicting pre-processing tasks and methods was shown earlier in *Sub-Section 2.2.1 (Figure 2)*.

The results of statistical analyses of the 15 month long pre-processed pressure signals (p_{REL}(t), p_{ATM}(t)) are shown below in *Figures 15, 16, 17, 18* for the PP1 pore pressure sensor.

Figure 15 shows p_{REL}(t) over the entire 15 month time span. Similarly to the previous *Section 4.1*, we analysed the cross-spectral gain of (p_{REL}(t), p_{ATM}(t)) at frequency 1/24h *(not shown here)*, and the wavelet component at nearly semi-diurnal time scale *(Figure 17)*. In addition, *Figure 16* shows the spectrum of Δp_{REL}(t), the first order difference of p_{REL}(t). Differencing enhances the peak frequencies, and allows one to detect a dominant 1/24h frequency, and a somewhat less dominant 1/12h frequency (subdominant but still visible). We see that these two characteristic frequencies can be detected in spite of the obvious non stationarity (large scale non linear trend) of the 15 month pore pressure signal *(Figure 15)*.

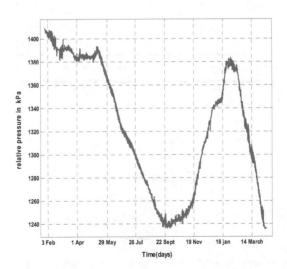

Fig. 15. Signal of the relative pore pressure PP1(t) (kPa) at Mont Terri Time span: 15 month (29/01/2004 to 12/04/2005). Time step of pre-processed signal: Δt = 30 min, and sampling step k=1 (i.e. no under-sampling).

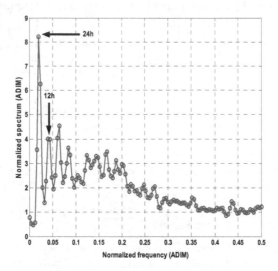

Fig. 16. Reduced spectrum of ΔPP1(t) (Δp_{REL}(t)), first difference of relative pore pressure PP1(t) [kPa] at Mont Terri. Time span ~15 months (29/01/2004 to12/04/2005). Time step of pre-processed signal: Δt=30 min (no sub-sampling).

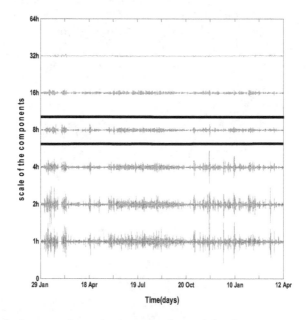

Fig. 17. Multi-resolution wavelet scale-time plot. Several dyadic components are shown here for the signal of relative pore pressure PP1(t) or p_{REL}(t) (pre-processed) at the Mont Terri site. Signal duration: from 29/01/2004 to 12/04/2005; time step Δt = 30 min; no sub-sampling (k=1).

Statistical Analyses of Pore Pressure Signals in Claystone During Excavation Works at the Mont
Terri Underground Research Laboratory

43

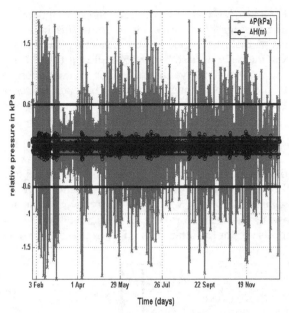

Fig. 18. Multi-resolution wavelet analysis: graph of the 8h dyadic scale component (near 12h) of the pre-processed relative pore pressure, PP1(t) or p_{RFI} (t). Record length ~15 months (from 29/01/2004 to 12/04/2005); Δt = 30 min, k=1 (no sub-sampling). The amplitude of semi-diurnal pressure fluctuations is $\Delta h \approx 0.06$ m equivalent water height.

The methods used here are again the same as those applied in the first example of *Section 4.1* where we analysed shorter, unprocessed pressure signals. We obtain here successively the specific storativity "Ss" and the effective elastic porosity "Φ". The final results are shown below for two different methods of quantifying the amplitude of semi-diurnal wavelet pressure fluctuations:

$|\Delta h|$ (*norm 1*) $\approx$ **0.06m**; **Ss** (*norm 1*) $\approx$ **3.3×10⁻⁷m⁻¹** Φ (*norm 1*) $\approx$ **1.0%**

$|\Delta h|$ (norm 2) $\approx$ **0.08m** ; **Ss** (norm 2) $\approx$ **2.6×10⁻⁷m⁻¹** Φ (*norm 2*) $\approx$ **0.9%**

where "*norm 1*" stands for the absolute mean, and "*norm 2*" stands for the Root Mean Square (or Standard Deviation).

5. Short time scale effects of gallery excavation on the evolution of pore pressure and of elastic properties (specific storativity Ss; elastic porosity)

5.1 Excavation of a gallery: observed effects on the evolution of pore pressure (syn-excavation phase, Mont Terri gallery Ga98)

We focus here on a total duration of ~5 months, corresponding to the syn-excavation period of gallery Ga98 (17th of November 1997 to 23rd of April 1998). The pre- and post-excavation periods were also analysed for comparison (*Fatmi 2009*), but it is sufficient to treat here the 5 month "syn-excavation" period. Furthermore, we will also focus, within this 5 month period, on a shorter phase of about 2 weeks of "fast" pressure changes (*see below*).

Upon inspection of *Figure 19*, it is clear that the pore pressure signal in sensor PP1 exhibits a strong response to the passage of the excavation front. This indicates the existence of a moving EdZ ("Excavation disturbed Zone") as the excavation front passes near the pressure sensors. Pressure rises gradually at first, then very rapidly ("fast" 11 day period). After this rise, pressure seems to reach a stable value but in fact, it starts a long and slow relaxation process (confirmed from analyses of longer records). The sharp rise of pore pressure can be explained as a hydro-mechanical disturbance of the geologic porous formation due to excavation works.

More precisely, *Figure 19* shows the pore pressure data from the PP1 sensor, in comparison with the evolving distance D(t) of the excavation front. The pre- and post-excavation phases are not shown. The 5 month "syn-excavation" phase was determined to run from the 17th of November 1997 to the 23rd of April 1998, and it is shown entirely in this figure (*see caption for other details and comments*). Clearly, within the 5 month syn-excavation phase, there is a remarkable sub-phase of "fast" pressure rise, lasting about two weeks only (~11 days). Similar effects were observed for the other sensor PP2 in the same borehole (*not shown here*).

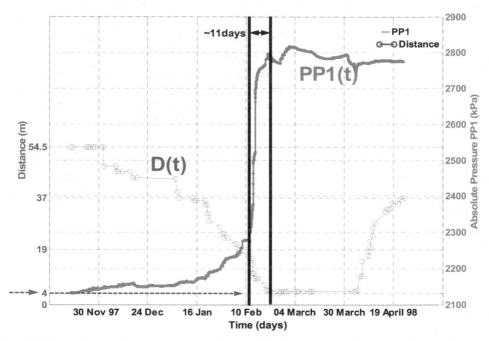

Fig. 19. Comparison of pore pressure and excavation front versus time over a 5 month period (syn-excavation). The green curve "PP1(t)" represents the time evolution of absolute pore pressure p(t) in sensor PP1. The blue curve with circles represents the distance D(t) between measurement chamber PP1 and the excavation front of gallery Ga98. The arrow at left indicates the minimal distance achieved (4 m), after which excavation was stopped for about a month (February/March 1998), before restarting around the 2nd week of April 1998. The sharp rise of pressure PP1(t) from 2250 to 2800 kPa ($\Delta p \approx +5.5$ bars) indicates the "fast" part of the syn-excavation phase, which lasted only 11 days (or roughly two weeks).

A study of the signal on longer time scales (*not shown in this text*) confirms in fact that the pore pressure disturbance was indeed reversible (hence the name "Excavation *disturbed* Zone", as explained earlier). Indeed, a slow relaxation of pore pressure was observed during a long period, somewhat less than two years, subsequent to the end of excavations works (there are almost two years counting from the start of excavation till the end of pressure relaxation: 30/11/1997-18/02/2000). The pore pressure at the end of this period reached approximately the value it had at the beginning of the period (pre-excavation) and it continued to decrease to reach finally the pre-pumping test value (*see Fatmi 2009, Fig.28, p.93, for more details*).

These observations justify our initial assumption and remark on the distinction between "EDZ" (*irreversibly damaged* "Excavation Damaged Zone"), and "EdZ" (*Excavation disturbed Zone*): the latter is thought to be a *reversibly disturbed* zone due to quasi-elastic hydro-mechanical variations of pressure and stress. These reversible variations occurred in the zone of our pressure sensor, and they caused the observed variations of pressure during the passage of the excavation front (which came within 4 m of the pressure sensor at the closest point).

Now, looking at the short and fast rise period of 11 days in **Figure 19**, it may seem "impossible" at first sight to detect a pattern of semi-diurnal pressure fluctuations due to earth tides, and even less, to follow possible changes in these fluctuations during the phase of rapid pressure change. Yet, although earth tides are partially masked, they can still be detected indirectly using proper filtering techniques. For example, applying differential filters can enhance subdominant frequencies, moving average filters can be used to look at residuals, and multi-resolution *wavelet* analysis can be used to extract precisely the semi-diurnal earth tide component (*as will be seen*).

Below, we will take advantage of the precise way in which multi-resolution wavelets can capture the semi-diurnal component of the original signal, and we will then analyse the new "12h" signal using envelope techniques, to finally obtain time-varying hydraulic parameters of the claystone during the passage of the excavation front nearby the pore pressure sensor.

5.2 Multi-resolution wavelet analysis of relative pressure signals during excavation: extraction of earth tide semi-diurnal component

First, we obtain the relative pore pressure $p_{REL}(t)$ from the absolute pore pressure signal $p(t)$ [*shown in* **Fig. 19** *above as* "$PP1(t)$"] by subtracting the atmospheric pressure signal $p_{ATM}(t)$ (*not shown*), whence: $p_{REL}(t) = p(t)-p_{ATM}(t)$.

Obviously, one should not expect $p_{ATM}(t)$ to be influenced by excavation works (rather, $p_{ATM}(t)$ contains barometric fluctuations, with peak frequency 1/24h, part of these having an impact on pore pressure).

The procedure to identify the elastic specific storativity Ss is then similar to that illustrated in **Section 4.1**, except that here, after extracting the semi-diurnal wavelet component "$C_{12H}(t)$" from the relative pore pressure signal $p_{REL}(t)$, we perform an additional analysis of the modulated amplitude, or "envelope" of $C_{12H}(t)$, in a statistical framework.

Note, from now on, the semi-diurnal wavelet component "$C_{12H}(t)$" is treated as a signal. The wavelet signal $C_{12H}(t)$ and its statistical envelope are now presented and analysed in the next section below (*Section 5.3*), in order to derive, from that information, the time-varying specific storativity "Ss(t)" during the syn-excavation phase.

5.3 Statistical envelope of semi-diurnal wavelet of relative pressure during excavation (Hilbert transform & method of maxima)

In this section, we present the wavelet component signal "$C_{12H}(t)$" of relative pore pressure, and we analyse its modulated amplitude by computing its Cramer-Leadbetter statistical envelope using MATLAB's Hilbert Transform (Section 3.5: Statistical envelope analysis...).

In addition, in some cases, we compare this theoretical envelope with an empirical one computed by our own "method of maxima". Briefly, the latter method generates an empirical envelope that has irregular time steps. The result of the empirical method agrees generally with the theoretical envelope except for isolated discrepancies (this is a minor point, but the interested reader can consult Fatmi 2009 for more details).

Figure 20 shows (in blue) the semi-diurnal signal "$C_{12H}(t)$", i.e. the 12h wavelet component of relative pressure in sensor PP1, and it shows also (superimposed in red) the statistical Cramer-Leadbetter envelope of "$C_{12H}(t)$" (envelope computed with the Hilbert Transform). Recall our gal is to analyse the effects of excavation of gallery Ga1998. Here, the entire "syn-excavation" phase of 5 months is shown (approximately from mid-november 1997 to mid-april 1998).

Figure 21 shows a temporal zoom of Figure 20 over a much shorter period. The time window is approximately one month, and it is centered around the "fast" sub-phase of syn-excavation, which lasts only 11 days.

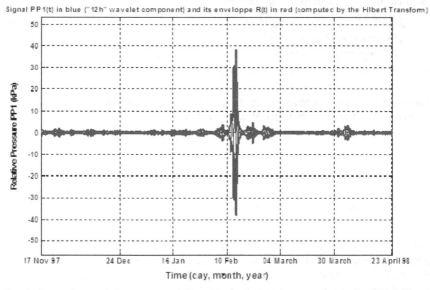

Fig. 20. Semi-diurnal wavelet component $C_{12H}(t)$ of relative pore pressure in PP1 (blue curve) and its envelope (red curve) during the "syn-excavation" phase of gallery Ga98 at Mont Terri (5 months, from 17 Nov 1997 to 23 April 1998, with Δt=30 mn time steps).

Statistical Analyses of Pore Pressure Signals in Claystone During Excavation Works at the Mont
Terri Underground Research Laboratory

47

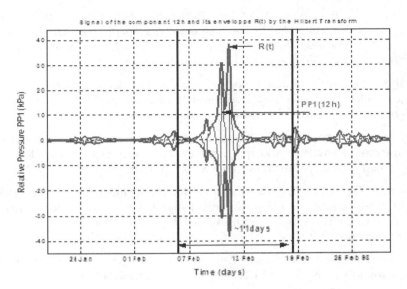

Fig. 21. Temporal zoom of the previous figure, shown over a time window of approximately 1 month (instead of 5 months in the previous figure). The (blue) curve designated as "PP1 (12h)" is the wavelet component signal $C_{12H}(t)$ extracted from PP1 relative pressure. The envelope is the (red) curve designated as "R(t)". The ~ 11 day period of fast pressure change occurs between the dates of 7th and 19th of February 1998.

Comparing *Figure 21* to *Figure 19*, it is now clear (and remarkable) that the 11 day phase of rapidly increasing pore pressure (*Figure 19*) corresponds to a significant increase of the *amplitude* of the semi-diurnal component of pore pressure (*Figure 21*). Recall, from *Fig. 19*, that this 11 day phase corresponds to the approach of the excavation front nearest to the PP1 pressure sensor (distance D(t) ~ 4 m at the closest point).

5.4 Calculation of specific storage Ss(t) as a function of time based on Bredehoeft's simplified hydro-mechanical model (strain / pressure)

By analogy with the procedure described in *Section 4.1-4.1.1*, we estimate again the elastic specific storativity Ss from the semi-diurnal wavelet component $C_{12H}(t)$ of relative pore pressure (using now the average of the 2 components at dyadic scales 8h and 16h, respectively). And again, we invoke the Bredehoeft earth tide strain/pressure model (see *Eq. 19* earlier).

Only, this time, instead of taking a constant amplitude for $C_{12H}(t)$, we use instead the time-modulated amplitude of $C_{12H}(t)$ represented by its statistical envelope "R(t)". On the other hand, since we still do not have direct *in situ* measurements of the volumetric strain $\Delta\varepsilon(t)$, we must replace $\Delta\varepsilon(t)$ by an indirect estimate of its amplitude based on data surveys: accordingly, we use again $|\Delta\varepsilon| = 2\times10^{-8}\ m^3/m^3$ (as in *Sections 4.1 and 4.2*). In this way, we obtain, instead of *Eq. 19*:

$$S_S(t) = \frac{|\Delta\varepsilon|}{R(t)}$$

(19b)

Figure 22 shows the time evolution of log₁₀ Ss(t), where Ss(t) was estimated using the empirical envelope R(t) computed by the method of maxima, rather than the theoretical Cramer-Leadbetter envelope computed by the Hilbert Transform. The evolution of the excavation front distance D(t) is also shown on the same graph.

Figure 23 shows the time evolution of log₁₀ Ss(t), using the theoretical Cramer-Leadbetter envelope (computed by the Hilbert Transform) to estimate Ss(t). The evolution of the excavation front distance D(t) is also shown on the same graph.

Comparing *Figure 22* and *Figure 23*, it is somewhat re-assuring that the two methods of estimating the wavelet envelope yield approximately the same results. Thus, we focus mostly on results obtained with the Hilbert Transform method for the remaining plots and analyses.

The time scale of *Figure 23* is about 1.5 month, including the interesting 11 day window of fast changes. It can be seen that the specific storativity *Ss(t)* decreases by one to two orders of magnitude during just *a few days*, as the excavation front approaches within roughly *6 m* to the PP1 pressure sensor (before reaching an even closer distance of *4 m*).

Figure 24 shows a zoomed view of Ss(t) with a cartesian ordinate rather than semi-log. The zoomed time window is only about 2 weeks or so, to see more details in the 11 day period of fast change.

Finally, *Figure 25* gives a broader picture, depicting the time evolution of log₁₀ Ss(t) over the entire "syn-excavation" phase of 5 months (including, again, the 11 day period of fast changes).

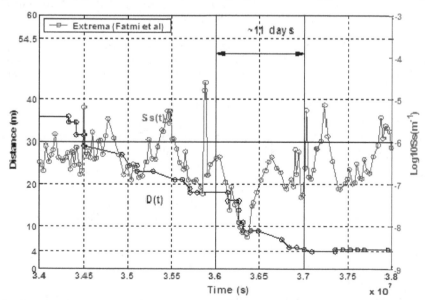

Fig. 22. Evolution of the decimal logarithm of Ss(t). The time window here is about 1.5 month or so, and it includes the 11 day period of fast change (*double arrow*). Here, Ss(t) is calculated from an empirical envelope (estimated with the lethod of maxima rather than the Hibert transform) of the semi-diurnal dyadic component of relative pore pressure PP1 during excavation (gallery Ga98, excavation from 17/11/1997 to 23/04/1998, time step Δt=30min).

Statistical Analyses of Pore Pressure Signals in Claystone During Excavation Works at the Mont
Terri Underground Research Laboratory

49

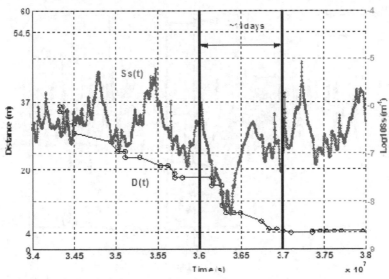

Fig. 23. Evolution of the decimal logarithm of Ss(t), on the same time window as the previous figure (1.5 month). Here, the red curve shows Ss(t) calculated from the theoretical envelope (Hilbert Transform, Cramer-Leadbetter envelope), applied to the semi-diurnal dyadic component of relative pore pressure PP1 during excavation of gallery Ga98. The distance D(t) of the excavation front from the PP1 pressure sensor is shown as the black curve with circles "o".

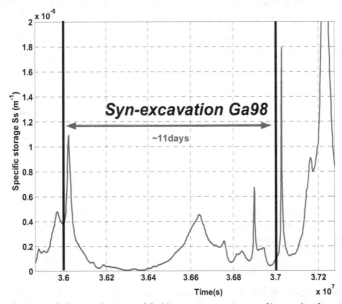

Fig. 24. Zoomed view of the evolution of Ss(t) on a cartesian ordinate (rather than logarithmic). The plot is centered on the 11 day period of fast pressure change, corresponding to the approach of the excavation front of gallery Ga98 to the pressure sensor.

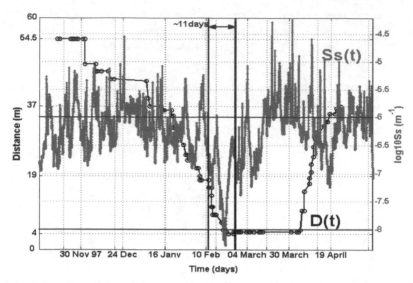

Fig. 25. A broader view of the evolution of log Ss(t) (red curve, computed with the Hilbert Transform envelope). Here the time window is the entire "syn-excavation" phase of 5 months (including, again, the 11 day period of very fast changes). The black curve with circles "o" represents the distance D(t) of the excavation front from the PP1 pressure sensor.

5.5 Calculation of elastic porosity Φ(t) from barometric effect (diurnal gain p_{REL}/p_{ATM}) and from the previously estimated Ss(t)

Finally, from the previously estimated evolution of Ss(t), and using the spectral barometric gain at 1/24h frequency (*as in Section 4.1.2*), it is possible to apply again the simplified hydro-mechanical model of *Eq. 21*, in order to estimate the time-varying elastic porosity Φ(t):

$$\Phi(t) = \frac{E_w \times B}{\rho \times g} \times S_S(t) \qquad (21b)$$

where *Ss(t)* is the time varying specific storativity obtained earlier from the strain/pressure envelope (**Eq.11b, Section 5.4**).

It should be observed that, in this analysis, the elastic porosity Φ(t) varies in time proportionally to Ss(t). It can be seen that a constant barometric efficiency "B" is assumed; the diurnal gain "g" between atmospheric and relative pore pressure is assumed constant during excavation. But, because pore pressure is influenced by excavation (even if air pressure is not), the assumption of constant barometric gain ("B" or "g") during excavation is debatable and should be verified.

In conclusion, a revised version of the procedure should be to analyse the diurnal barometric gain, like the semi-diurnal earth tide effects, in a wavelet framework rather than in a Fourier spectral framework. This would provide a tool to either capture or eliminate, depending on purpose, any non stationary / evolutionary phenomena.

Statistical Analyses of Pore Pressure Signals in Claystone During Excavation Works at the Mont
Terri Underground Research Laboratory

51

6. Summary, conclusions, and outlook

6.1 Summary and conclusions

We have presented in this chapter a set of statistical methods for pre-processing and analysing multivariate hydro-geo-meteorological time series, and we have developed hydro-mechanical interpretations, in order to infer hydro-geological properties such as Ss and Φ. The aim of this study is to quantify some of the properties of the clay rock in terms of elastic response to barometric and earth tides fluctuations, both *globally* and *locally* in space-time. We focused on atmospheric and pore water pressure signals measured in Opalinus clay stone, at the Mont Terri underground rock laboratory (URL), in the framework of Mont Terri's LP 14 experiment.

Pre-processing methods were developed in order to detect and correct some defects common to most time series data, like aberrations or outliers, variable time steps, and missing data. These pre-processing steps are necessary, in most cases, to obtain signals that are both "clean" and long enough for useful analyses. Durations of a couple of weeks to a few months are needed for short scale analyses related to excavation works. Ideally, several years are required for global characterizations, and even more for climate trend analyses.

Once the pressure signals were pre-processed (reconstructed), statistical analyses were performed with cross-correlation functions and cross-spectra, and with multi-resolution wavelet analyses. These techniques were then used to determine "globally" some hydro-mechanical parameters of the Opalinus clay: elastic specific storativity (Ss), and effective elastic porosity (Φ). The values compare favourably with data from the literature, obtained independently with completely different methods. In particular, the simplified hydro-mechanical model of *Bredehoeft* relating $p_{REL}(t)$ to earth tides was applied to the "12h" dyadic component of $p_{REL}(t)$, leading to a global estimated Ss value of 10^{-6} m^{-1} (1E-6 m^{-1}) in excellent agreement with that deduced from hydraulic tests at the Mont Terri site.

After this "global" characterization of the clay from long term pore pressure measurements, we turned our attention to the (significant) impact of excavation works on pore pressure signals. We devised an original method to infer the time evolution of clay rock properties (Ss(t)) due to the excavation disturbance. The pressure disturbance coïncides (with some lead time or negative delay) with the arrival near the pressure sensor of the excavation front D(t) (or of the "EdZ" around and ahead of the front).

Indeed, pressure signals recorded in the PP1 chamber of the BPP-1 borehole during excavation of the Ga98 gallery, have shown a net increase of absolute pressure (+5.5 bars) over the 11 day period of fast change (or as much as +7 bars over the total 5 month "*syn-excavation*" phase). This is evidence of a direct hydro-mechanical disturbance of the rock during the passage of the excavation front.

A study of the signal on longer time scales (from pre- to post-excavation stages, *not shown in this text for lack of space*) confirmed that the disturbance due to excavation was indeed reversible. A slow relaxation of pore pressure was observed during a period of about two years (or somewhat less) subsequent to the end of excavation, reaching approximately the value it had before excavation. This observation appears to justify our initial remark on the distinction between "EDZ" (*irreversibly damaged* "Excavation Damaged Zone"), and "EdZ" (*Excavation disturbed Zone*). The latter "EdZ" is a *reversibly disturbed* zone due to quasi-elastic hydro-mechanical variations of pressure and stress.

In summary, the signal analysis and processing methods applied at various (long/short) time scales, exploit the influence of natural "forcings" like earth tides and barometric fluctuations on pore pressure signals. Combining multi-resolution wavelet and he Cramer/Hilbert envelope analysis reveals a time modulation (disturbance) of the "12h" dyadic wavelet component, suggesting a modification in time of the specific storativity $Ss(t)$ during the excavation of gallery passing nearby the pressure sensor. During the passage of excavation, there is a net decrease of $Ss(t)$ by almost two orders of magnitude, from 10^{-6} (1E-6) m⁻¹, down to 10^{-8} (1E-8) m⁻¹. This minimum occurs during a short period of just a few days, corresponding to front crossing (actually, in advance of front crossing by a couple of meters).

This result demonstrates how statistical analyses on pore pressures can be a valuable tool in the assessment of potential clay repositories for the purpose of high level radioactive waste disposal in deep, isolating geologic formations.

6.2 Outlook

A useful extension of this work would involve the development of a multivariate or "*multi-cross* analysis" of time series data $\{Xi(t), i = 1, 2, ..., K\}$, where the Xi's are either reconstructed or raw signals, involving possibly a number of pore pressures measured in different boreholes at different measurement sections, as well as other types of signals such as air pressure, air temperature, air humidity (relative or absolute), wall crack displacements, and moisture contents in the rock.

Accordingly, in multi-cross analysis, the results would be expressed in matrix form $M_{XiXj}(s)$ where "s" is the relevant parameter (time lag, frequency, etc.): thus $M_{XiXj}(s)$ may represent a K×K matrix of cross-correlation functions, or of Fourier cross-spectra, and/or, of multi-resolution cross-wavelets (or their cross-correlations at given scales). These analyses will be useful for assessing the added value of *all* signals collected at the site, in terms of the "information" gained concerning the hydro-geologic properties of clay rock at different time scales (hourly, daily, yearly).

The purpose of such multi-cross analyses might be to integrate the complete set of all available signals in order to solve an *inverse problem*, namely, the identification of hydro-geologic properties from observations of state variables. Thus, in the work presented here, the state variables were the pore pressure at different sensors, atmospheric pressure, and (indirectly) tidal strain. The properties to be identified were (Ss, Φ) - but this set of unknown coefficients could be extended in future, in a generalized treatment of the "hydro-geologic identification problem".

For instance, techniques to estimate the hydraulic conductivity "K" from natural signals sampled at different positions might be developed as a complement to classical well pumping tests (*e.g. pulse tests*). Indeed "K" is an important property for a geologic radioactive waste repository, as "K" must be low to efficiently isolate radioactive waste from the environment. Furthermore, it is well known that "K" may vary greatly in geologic media, depending on the spatial scale of analysis (e.g. *Ababou 2008, Sahimi 1993, Gelhar 1993, Ababou et al. 1990,*). With the "multi-cross" concept, the spatial scales of analysis can be associated with the distances between sensors.

The multi-cross analysis of these signals will also serve as a complement to physically based models, with some of the signals considered as inputs, and others considered as outputs of the models. It will be very useful to link the statistical approaches with physically-based models that can account for the hydro-mechanical coupling between pore pressure and *exogeneous* fluctuations (piezometric variations in adjacent aquifers, barometric fluctuations, and earth tides).

In addition, some aspects of the procedures developed in this work could be modified in future:

- Thus, we developed a sequential procedure to estimate first "Ss" then "Φ". First, we estimated "Ss" from wavelet analysis of semi-diurnal earth tide effects, and secondly, we estimated "Φ" from spectral analysis of barometric gain at diurnal frequency (given the previously estimated "Ss"). In reality, these 2 steps need not be implemented sequentially: they can be viewed as a system of 2 equations with 2 unknowns (Ss, Φ) to be determined both at once.

- To estimate the global "Ss" and (also) the evolutionary "Ss(t)" during excavation, we used a fixed value, $|\Delta\varepsilon| = 2 \times 10^{-8}$ m^3/m^3, for the amplitude of the volumetric strain related to the M2 semi-diurnal earth tide (*based on Melchior 1978, Bredhoeft 1967, and others*). This is because the site specific signal of volumetric strain $\Delta\varepsilon(t)$ was not available, so we had to estimate its amplitude indirectly from data surveys. Nevertheless, our method can be extended to the case where the strain signal $\Delta\varepsilon(t)$ is available. The new procedure would be to apply semi-diurnal wavelet analysis, then envelope analysis, to the $\Delta\varepsilon(t)$ signal, in a way similar to that already implemented for relative pore pressure $p_{REL}(t)$.

Another extension worth considering concerns the simplified hydromechanical "models" which we have used so far for interpreting the pressure signals. These models need to be enhanced:

- Thus, it was noted that "Ss" represents an equivalent elastic coefficient for the water filled poro-elastic rock. However, this concept rests on a simplified view of poro-elastic deformations, usually applied to confined aquifers. Its interpretation for the hydrogeologic configuration and boundary conditions of the claystone at Mont Terri is not straightforward and requires further analyses.

- Also, a more "physical" hydromechanical approach based on Biot's poro-elastic theory (*Biot 1956*) could be developed, as in the coupled T-H-M models developed in *Ababou et al. (2005)* and *Canamon (2009)*. Such models should be upgraded with explicit coupling of pore pressure to external "forcing" functions (air pressure, earth tides, etc.), in order to better interpret the effects of these signals.

- Finally, the hydro-mechanical model could be constructed to predict space-time propagations explicitly. Thus, *Mallet et al. (2007, 2008)* developed a simple model to illustrate pore pressure fluctuations and propagation under boundary forcing in a compressible hydrogeologic medium.

In summary, the simplified hydro-mechanical models used in this study can be replaced, in future, with models based on PDE systems. The statistical "interpretations" of pressure signals presented in this text will then appear as the solution of PDE inverse problems (in a statistical setting). This point of view sheds a new light on the coefficients (Ss, Φ) obtained in the present study. They can be viewed as the solution of an inverse problem based on spectral gain (p_{REL}/p_{ATM}) and on wavelet amplitude ratio ($|\Delta\epsilon|/|\Delta p_{REL}|$). For this reason, it is thought that the statistical methods used here can be adapted for solving an inverse problem involving a hydro-mechanical PDE model forced by pressure signals and other measurements (humidity, temperature, etc.).

7. Acknowledgments

This chapter presents data analyses developed within Mont Terri's LP 14 experiment (Long term Pressure monitoring project, phase 14), as part of the international research programs of the MONT TERRI CONSORTIUM on the properties of claystone as a potential geologic repository for radioactive waste. The authors acknowledge the support of institutions participating in the LP 14 project at Mont Terri (ANDRA, NAGRA, IRSN, NWMO, SwissTopo). In addition, the authors gratefully acknowledge the help and advice provided by the following persons: Paul Bossart (SwissTopo); Martin Cruchaudet (ANDRA); David Jaeggi (SwissTopo); Herwig Mueller (NAGRA); Eric Sykes (NWMO). The authors wish also to thank Alain Mangin (CNRS-UPS Toulouse, Laboratoire Souterrain de Moulis) who initiated some of the statistical tools and methods used in this project, and others who provided scientific advice on various issues (cracks, geophysics, signals, hydrogeology, porous media): Andrea Möri (SwissTopo); Jacques Delay (ANDRA, LSM/HM Bure); Israel Cañamón Valera (Universidad Politecnica de Madrid); Moumtaz Razak (Université de Poitiers, HYDRASA); and Michel Quintard (IMFT, Toulouse).

8. Appendix: Symbols and abbreviations

ABBREVIATIONS, ACRONYMS

ANDRA	Agence Nationale pour la gestion des Déchets RAdioactifs (*France*)
BPP	Borehole instrumented for Pore Pressure measurements (*Mont Terri site*)
EdZ	Excavation disturbed Zone (reversibly disturbed zone around an excavation in rock, elastic hydro-mechanical disturbance)
EDZ	Excavation Damaged Zone (irreversibly damaged, fissured and fractured zone around an excavation in rock)
IRSN	Institut de Radioprotection et de Sûreté Nucléaire (France)
LP14	Long term Pressure monitoring project - phase 14 (Mont Terri Project)
NAGRA	National Cooperative for the Disposal of Radioactive Waste (*Switzerland*)
NWMO	Nuclear Waste Management Organization (*Canada*)
PP	Pore Pressure, i.e., absolute pore water pressure (*Mont Terri site*)
SwissTopo	Swiss Federal Office of Topography (*and Mont Terri Project*)
T-H-M or THM	Thermo-Hydro-Mechanical (coupled models)
URL	Underground Rock Laboratory (*or:* Underground Research Laboratory)

Statistical Analyses of Pore Pressure Signals in Claystone During Excavation Works at the Mont
Terri Underground Research Laboratory

55

MATHEMATICAL SYMBOLS: LATIN

$A_m(t)$	Wavelet approximation of order "m" (multi-resolution dyadic scale "m")
B, BE, B_E	Barometric efficiency, [dimensionless] or [Pa/Pa] (relative pressure / atmospheric pressure effects)
$C_{12H}(t)$	Wavelet component of scale 12h (semi-diurnal)
$C_m(t)$	Wavelet component of dyadic scale "m" (time scale = $2^m \square \square t$)
$C_{XY}(\tau)$	Cross-covariance function of X(t) and Y(t), versus time-lag [units of Cxy] = [units of X(t)] $\square$ [units of Y(t)]
D(j) or D(τ(j))	Tuckey-2 filter function of time-lag (used for estimation of Fourier spectra)
$D_m(t)$	Wavelet "detail" at dyadic scale "m" (*see also: wavelet component $C_m(t)$*)
f	Frequency "f" in [Hz] or [Number of cycles / second]; "f" is related to angular frequency "ω" by: f = ω /2π.
$g_{XY}(f)$	Reduced spectral gain *vs.* frequency, obtained from cross-spectrum. The complex-valued reduced gain is defined as: $g_{XY}(f) = s_{XY}(f)/s_{XX}(f)$.
H	Total hydraulic head [m], related to relative pressure by: $\rho g \delta H = \delta p_{REL}$
k	Permeability (intrinsic Darcy permeability) in [m²]
K	Hydraulic conductivity [m/s] (sometimes unduly called "permeability")
$M2$, M_2	Moon semi-diurnal component of astronomic tides (here, earth tides)
p, p_{ABS}	Pore pressure (or "absolute" pore pressure), [Pa]. (*)
p_{ATM}	Atmospheric pressure (or air pressure), [Pa]. (*)
p_{REL}	Relative pore pressure [Pa], defined as : $p_{REL} = p - p_{ATM}$ (*)
$R(t)$	Statistical envelope of a signal (the envelope is in fact ±R(t)).
$R_{XY}(\tau)$	Cross-correlation function of X(t) and Y(t), versus time-lag; Rxy is the normalized version of Cxy: [units of Rxy] = [dimensionless]
$s_{XY}(f)$	Reduced cross-spectrum of signals X(t) and Y(t), *vs.* frequency "f". It is equal to the cross-spectrum $S_{XY}(f)$ divided by $(\sigma_X \times \sigma_Y)$. The cross-spectrum is complex-valued (even if the signals are real-valued).

$s_{XX}(f)$ Reduced auto-spectrum of X(t) *vs.* frequency "f";

it is equal to the spectrum $S_{XX}(f)$ divided by the variance σ_X^2.

The (auto)-spectrum is always real-valued for a real-valued process X(t).

S_S Specific elastic storativity, [m^{-1}] or [m^3/m^3 *per meter of head* δH]

(*) Pressures "p" are sometimes denoted with a capital "P" within the text.

MATHEMATICAL SYMBOLS: GREEK

$\Delta \varepsilon, \Delta \varepsilon_{TIDAL}$ Tidal dilatation, volumetric strain [m^3/m^3]
(here, this notation is used for the *amplitude* of the tidal strain).

Δt Time step

τ Time-lag or "delay", used in correlation function analyses:
if t' and t" are two time instants, the time lag is τ = t"-t'.

ω Angular frequency, or "pulsation", in [radians/s];
"ω" is related to frequency "f" [Hz] by : ω = 2πf.

Φ Effective elastic porosity [m^3/m^3]

9. References

Ababou R. (2008): *Quantitative Stochastic Hydrogeology: the Heterogeneous Environment*. Chap. 8 in Part III of "Overexploitation & Contamination of Shared Groundwater Resources: Management, (Bio)Technological, and Political Approaches to Avoid Conflicts." NATO-ASI: Advanced Studies Institute Series, C.J.G. Darnault (ed.), Springer Science & Business Media BV, January 2008 (pp. 119-182).

Ababou R., I. Cañamón, F. J. Elorza (2005): Thermo-Hydro-Mechanical simulation of a 3D fractured porous rock: preliminary study of coupled matrix-fracture hydraulics. Proceedings, *FEMLAB 2005 Conference: "Une conférence multiphysique"*. Paris, France, 15 November 2005, pp.193-198.

Ababou R., L. W. Gelhar (1990): *Self-Similar Randomness and Spectral Conditioning: Analysis of Scale Effects in Subsurface Hydrology*. Chap. XIV in "Dynamics of Fluids in Hierarchical Porous Media", J. Cushman (editor), Academic Press, New York, 1990 (pp. 393-428).

Ababou R., L.W. Gelhar, D. McLaughlin (1988): *Three-Dimensional Flow in Random Porous Media*. Tech. Report No. 318, Ralph Parsons Laboratory for Water Resources & Hydrodynamics, Department of Civil Engineering, Massachusetts Institute of Technology (MIT), Cambridge, Massachusetts, USA. (2 vols., 833 pp., March 1988). [See also *Ababou R., 1988, PhD thesis, MIT* :
http://libraries.mit.edu/ http://dspace.mit.edu/handle/1721.1/14675].

Biot M.A. (1956): General Solutions of the Equations of Elasticity and Consolidation for a Porous Material. *Journal of Applied Mechanics* 23: pp. 91-96. 1956.

Statistical Analyses of Pore Pressure Signals in Claystone During Excavation Works at the Mont
Terri Underground Research Laboratory

57

Blackman R.B., Tukey J.W. (1958): *The Measurement of Power Spectra.* Dover Publications. NY, 1958.

Bossart P., Meier P.M., Moeri A., Trick T., and Mayor J.-C. (2002). Geological and hydraulic characterisation of the excavation disturbed zone in the Opalinus Clay of the Mont Terri Rock Laboratory. *Engineering Geology* 66:19-38.

Blümling P., Frederic B., Patrick L., Martin C.-D. (2007). The excavation damaged zone in clay formations time-dependent behaviour and influence on performance assessment. *J. Phys. Chem. Earth* 32:588–599.

Box, G.E.P., G.M. Jenkins (1976): *Time Series Analysis, Forecasting, and Control.* Revised Edition. San Francisco, CA: Holden-Day Publishers.

Bras R., I. Rodriguez-Iturbe (1985): Random Functions in Hydrology. Dover, NY, 1985 (re-edited: 1993).

Bredehoeft J.D. (1967): *Response of Well-aquifer Systems to Earth Tides.* U.S Geological Survey, Washington, D.C., 20242.

Cañamón I., F.J. Elorza, A.Mangin, P.L. Martin, R. Rodriguez (2004): Wavelets and statistical techniques for data analysis in a mock-up high-level waste storage experiment. *International Journal of Wavelets, Multiresolution and Information Processing (IJWMIP):* Volume: 2, Issue: 4 (2004), pp. 351-370. DOI: 10.1142/S0219691304000585.

Cañamón Valera I., 2009: *Coupled phenomena in 3D fractured media (analysis and modeling of thermo-hydro-mechanical processes).* VDM Verlag Dr. Müller, 2009, 212 pp. (ISBN: 3639189973, 978-3-639-18997-1).

Cañamón Valera I., 2006: *Analysis and Modeling of Coupled Thermo-Hydro-Mechanical Phenomena in 3D Fractured Media (Analyse et Modélisation des Phénomènes Couplés Thermo-Hydro-Mécaniques en Milieux Fracturés 3D).* Thèse de doctorat (en anglais). IMFT - Institut National Polytech. de Toulouse & Universidad Politécnica de Madrid (Escuela Técnica Superior de Ingenieros de Minas). Thesis supervisors: R. Ababou (INPT-IMFT), J. Elorza (UPM-DMAMI).

Daubechies I. (1988): Orthonormal bases of compactly supported wavelets. *Comm. Pure & Appl. Math.,* XLI, pp. 909-996, 1988.

Daubechies I. (1991): "Ten lectures on wavelets". *CBMS-NSF Series Appl. Math.,* SIAM, 1991.

Fatmi H., R. Ababou, J.-M. Matray (2008):"Statistical pre-processing and analyses of hydrogeo-meteorological time series in a geologic clay site (methodology and first results for Mont Terri's PP experiment)". *Journal of Physics & Chemistry of the Earth (JPCE),* Special Issue «Clays in Natural & Engineered Barriers for Radioactive Waste Confinement» (CLAY'2007): 33(2008) S14-S23.

Fatmi, H., Ababou R., Matray J.M., Joly C. (2010) "Hydrogeologic characterization and evolution of the 'Excavation Damaged Zone' by statistical analyses of pressure signals: Application to galleries excavated at the claystone sites of Mont Terri (Ga98) and Tournemire (Ga03)". *4th Internat. Meet. "Clays in Natural & Eng. Barriers for Rad. Waste Confin.",* Nantes 2010. Poster P/EDZ/CH/03: Book of abstracts: pp. 809-810.

Fatmi H. (2009). *Méthodologie d'analyse des signaux et caractérisation hydrogéologique: Application aux chroniques de données obtenues aux laboratoires souterrains du Mont Terri, Tournemire et Meuse/Haute-Marne.* Ph.D. thesis, Institut National Polytechnique de Toulouse, Toulouse & Fontenay-aux-Roses (France), June 2009, 248 pp., in french. [*Available on line at* INPT].

Fatmi H., R. Ababou, J.-M. Matray, Ch. Nussbaum (2011): "Statistical analyses of pressure signals, hydrogeologic characterization and evolution of *Excavation Damaged Zone* (claystone sites of Mont Terri and Tournemire)." *Proceedings MAMERN11*: 4th Internat. Conf. Approx. Methods and Numer. Modelling in Envir. & Natural Resources, Saidia (Morocco), May 23-26, 2011. B.Amaziane, D.Barrera, H.Mraoui, M.L.Rodriguez & D.Sbibih (eds.), Univ. de Granada, ISBN:078-84-338-5230-4 (2011), pp.325-329.

Fatmi H., R Ababou, J.-M. Matray (2007) : "Méthodologie de prétraitement et d'analyse du signal. Application aux chroniques de données multivariées obtenues au Mont Terri". Rapport IRSN DEI/SARG/n°2007-035 (rapport d'avancement No.1 pour 2006/07), Fontenay-aux-Roses, 27 pp., 2007.

Freeze R.A. & J.A. Cherry (1979): *Groundwater*. Prentice Hall, Englewood Cliffs NJ, 604 pp.

Gelhar L.W. (1993): *Stochastic Subsurface Hydrology*. Prentice Hall, Englewood Cliffs, New Jersey, 390 pp., 1993.

Hsieh P.A., Brededhoeft J.D., Rojstaczer A. (1988). Response of well aquifer to earth tides: problem revisited. *Water Resources Research*, 24, 3, pp 468-472.

Huang N. E., Zheng Shen, Long S. R. (1999): A new view of nonlinear water waves: The Hilbert spectrum. *Annual Review of Fluid Mechanics*, Vol. 31, pp. 417-457 (2p.3/4), 1999. ISSN:0066-4189.

Jacob C.E. (1940): On the flow of water in an artesian aquifer. Transactions of the American Geophysical Union, 2, pp. 574-786, 1940.

Labat D. (2005): Recent advances in wavelet analyses: Part 1. A review of concepts. *Journal of Hydrology*: 314 (2005) 275-288.

Labat D., R. Ababou, A. Mangin, 1999a: Wavelet analysis in karstic hydrology. 1rst part: Univariate analysis of rainfall rates and karstic spring runoffs. *C.R.Acad.Sci. de Paris*, Sci. de la Terre et des Planètes (Earth & Planetary Sciences), 1999, 329, pp.873-879.

Labat D., R. Ababou, A. Mangin, 1999b: Wavelet analysis in karstic hydrology. 2nd part: Rainfall-runoff cross-wavelet analysis. *C.R. Acad.Sci. de Paris*, Sciences Terre et Planètes (Earth & Planetary Sciences), 1999, 329, pp.881-887.

Labat D., R. Ababou, A. Mangin (1999c): Linear and Nonlinear Models Accuracy in Karstic Springflow Prediction at Different Time Scales. *SERRA - Stochastic Environmental Research & Risk Assessment*, 13(1999):337-364, Springer-Verlag.

Labat, R. Ababou, A. Mangin (2000a): Rainfall-runoff relations for karstic springs – Part I : Convolution and spectral analyses. *Journal of Hydrology*, Vol. (238), Issues 3-4 (5 Dec.2000): pp. 123-148.

Labat D., R. Ababou, A. Mangin (2000b): Rainfall-runoff relations for karstic springs - Part II : Continuous wavelet and discrete orthogonal multiresolution analyses. *Journal of Hydrology*, Vol. (238), Issues 3-4 (5 Dec.2000): pp. 149-178.

Labat D., R. Ababou, A. Mangin (2002): Analyse multirésolution croisée de pluies et débits de sources karstiques. *C. R. Geosciences* (Elsevier), 334 (2002) 551-556.

Mallat, S. (1989): A Theory For Multiresolution Signal Decomposition: The Wavelet Representation. *IEEE Trans. on Pattern Anal. & Mach. Int.*, 11(7), pp. 674-693, 1989.

Mallet, A., R. Ababou, J.-M. Matray (2007): Multidimensional modeling of a poro-elastic medium: identification of the behavior of a clay formation from a stochastic representation of piezometric fluctuations. Poster Abstract, Book of Abstracts, *Internat. Conf. CLAY 2007*, Lille, France, 2007.

Mallet A., J.-M. Matray, R. Ababou (2008) : *Etude des fluctuations piézométriques à l'intérieur d'un milieu poreux saturé, et détermination des paramètres caractéristiques du milieu poreux*. Note Technique IRSN/DEI/SARG/2008-024 (Version 1) - Technical Note of the Institut de Radioprotection et Sûreté Nucléaire (IRSN), Fontenay-aux-Roses, France, 2008.

Mangin A. (1984). Pour une meilleure connaissance des systèmes hydrologiques à partir des analyses corrélatoires et spectrales. *J. Hydrol.*, 67, 25-43, 1984.

Marsaud B., A. Mangin, F. Belc (1993): Estimation des caractéristiques physiques d'aquifères profonds à partir de l'incidence barométrique et des marées terrestres. *Journal of Hydrology*, 144 (1993) 85-100.

Marschall P., Ababou R., Bossart P., Cruchaudet M., Fatmi H., Matray J.-M., T. Tanaka & P. Vogel (2009). Chap. 8: Hydrogeology experiments. In: *"Report of the Swiss Geological Survey N°3: Mont Terri Rock Laboratory-Project, Program 1996 to 2007 & Results"*. Bossart P. & Thury M. (eds.), Federal Office of Topography - SwissTopo (2008/09), ISBN 978-3-302-40016-7, Wabern, Swizerland, pp. 95-106.

Massmann J. (2009). *Modelling of Excavation Induced Coupled Hydraulic-Mechanical Processes in Claystone*. Dissertation, Institut ffir Strômungsmchanik and Elektron. Rechnen im Bauwesen der Leibniz Univ. Hanover, Hanover, Germany.

Max J., (1980) : *Méthodes et techniques de traitement du signal et application aux mesures physiques*, Tome I (354 pp.) et Tome II (454 pp.), Chap.I-XXVIII, Masson, Paris.

Max J., J.L. Lacoume, (1996) : *Méthodes et techniques de traitement du signal*. Dunod, 5ème ed., 355 pp.

Melchior P. (1978): *The tides of the planet earth*. Pergamon Press, Paris, 609 pp.

Möri A., Bossart P., Matray J.-M., Müller H., Frank H, & Ababou R. & Fatmi H. (2012): "Mont Terri project, cyclic deformations in the Opalinus clay". Special Issue "Clays in Natural & Engineered Barriers for Radioactive Waste Confinement" (4th Internat. Meeting Clays 2010, Nantes, 29 March – 1rst April 2010). *Journal of Physics & Chemistry of the Earth* (ISSN 1474-7065, DOI: ---); *in press (2012)*.

Nussbaum C., Bossart P., Amann F. and Aubourg C. (2011). Analysis of tectonic structures and excavation induced fractures in the Opalinus Clay, Mont Terri underground rock laboratory (Switzerland). *Swiss Journal of Geosciences*: Volume 104, Issue 2 (2011), Pages 187-210. DOI: 10.1007/s00015-011-0070-4.

Nussbaum, C., Wileveau, Y., Bossart, P., Möri, A., Armand, G., (2007). "Why are the geometries of the EDZ fracture networks different in the Mont Terri and Meuse/Haute-Marne Rock Laboratories? Structural Approach." *CLAY'2007: Clays In Natural & Engineered Barriers for Radioactive Waste Confinement*, Lille, France, 17-18 Sept. 2007. (2 pp., extended abstract).

Papoulis A., S.U. Pillai (2002): *Probability, Random Variables, and Stochastic Processes*. Mc Graw-Hill Book Company, New York. 2002 (4th ed, 16 Chapters, 852 pp.).

Priestley M.B. (1981). *Spectral analysis and time series*. Academic Press, 890 pp.

Sahimi M. (1993): Flow phenomena in rocks: from continuum models to fractals, percolation, cellular automata, and simulated annealing. Reviews of Modern Physics, Vol. 65, N°4, pp. 1393-1533 (1993).

Schaeren, G.and Norbert, J. (1989): Tunnels du Mont Terri et du Mont Russelin. La traversée des "roches à risques": marnes et marnes à anhydrite. *Soc. Suisse Ing. Arch.*, Doc. SIA D 037, 19–24.

Takeuchi, H (1950): On the earth tide of the compressible earth of variable density and elasticity. *Transactions American Geophysical Union*: 31, 651-689, 1950.

Terzaghi, V.K. (1936): "The Shearing Resistance of Saturated Soils and the Angle between the Planes of Shear". First International Conference of Soil Mechanics, Vol. 1: pp. 54-56. Harvard University. 1936.

Thury M. & Bossart P. (1999): "Mont Terri rock laboratory. Results of hydrogeological, Geochemical and geothechnical Experiments Performed in 1996 and 1997". *Swiss National Hydrological and Geological Survey*. Geologic Report N°23. Bern, Switzerland, 1999.

Vanmarke E. (1983): *Random Fields Analysis and Synthesis*. The MIT Press, Cambridge MA, USA, 382 p.

Veneziano D. (1979): Envelopes of Vector Random processes and their Crossing Rates. *The Annals of Probability*: 1979, Vol.7, No.1, 62-74.

Wang Y., R. Ababou, M. Marcoux (2010): "Signal Processing of Water Level Fluctuations in a Sloping Sandy Beach Modeled in a Laboratory Wave Canal", in: *Experimental & Applied Modeling of Unsaturated Soils GSP202* (Proc.2010 GeoShanghai International Conference, Part II: Applied Modeling and Analyses. L. R. Hoyos, X. Zhang, A. J. Puppala, eds.). American Society of Civil Engineers: *ASCE Geotech. Special Publi. (GSP)*: Vol.376, No.41103, pp.204-210. http://dx.doi.org/10.1061/41103(376)26

Yaglom, A. M. (1987). *Correlation Theory of Stationary and Related Random Functions: Basic Results*. Springer-Verlag, New York.

Yevjevich V., (1972): *Stochastic Processes in Hydrology*. Water Resources Publications, Fort Collins, Colorado, USA, 8 Chapters, 276 pp.

2

Hydrogeologic Characterization of Fractured Rock Masses Intended for Disposal of Radioactive Waste

Donald M. Reeves, Rishi Parashar and Yong Zhang
Desert Research Institute
USA

1. Introduction

There are currently 441 nuclear power reactors in operation or under construction distributed over 30 countries (International Atomic Energy Agency, 2011). The global radioactive waste inventory reported as storage in 2008 was approximately 17.6 million cubic meters: 21% short-lived, low- and intermediate-level waste, 77% long-lived, low- and intermediate-level waste and 2% high-level waste (International Atomic Energy Agency, 2011). There is a consensus among most of the scientific community that geologic repositories offer the best solution for the long-term disposal of radioactive waste. In the United States, for example, geologic disposal is considered the only technically feasible, long-term strategy for isolating radioactive waste from the biosphere without active management (Long & Ewing, 2004; National Research Council, 2001; Nuclear Energy Agency, 1999).

The process of selecting geologic repositories is an issue for many countries with radioactive waste, and several countries including Belgium, Canada, Finland, France, Japan, Republic of Korea, Sweden, Switzerland and the United States have underground research laboratories to conduct in-situ tests related to radioactive waste disposal (International Atomic Energy Agency, 2001). The Swedish Nuclear Fuel and Waste Management Company (SKB) recently selected Östhammar, Sweden as the final spent fuel geologic repository site after a nearly 20 year selection process. Once operational, this facility will be the first repository in the world designated for long-term disposal of high-level radioactive waste.

Repositories intended for long-term storage of high-level radioactive waste are comprised of both engineered and geologic barriers to either isolate from or impede the release of radioactive elements (radionuclides) to the biosphere. The engineered barrier serves as the primary barrier to radionuclide transport and describes systems that consist of (a) form of the waste, such as radionuclides contained within a solid, vitric matrix rather than in an aqueous phase, (b) waste canister, (c) backfill and buffer, and (d) tunnel grouting. In the event of radionuclides circumventing the engineered barrier, the geologic barrier serves as a secondary impediment to the release of radionuclides to the biosphere. The geologic barrier relies on the intrinsic ability of the host geologic medium to limit the transport of radionuclides.

The intrinsic ability of the host rock to limit radionuclide migration is a complex interplay between advective-dispersive motion of radionuclides in dissolved and colloidal form in groundwater, and retention processes such as adsorption and molecular diffusion into low velocity zones. These processes are present for both unsaturated and saturated hydrologic conditions. Thus, hydrogeologic characterization of a potential repository must provide estimates of advective transport velocity based on permeability and porosity of the host rock; delineate the surrounding groundwater flow system including regional flow directions, hydraulic gradients, and zones of recharge and discharge; and determine the potential significance of radionuclide retention mechanisms.

In this Chapter, we focus our attention on the hydrogeologic characterization of fractured rock masses intended for the disposal of radioactive waste. The emphasis on fractured rock masses is two-fold. First, many repository sites proposed for high-level radioactive waste disposal are comprised of low-permeability hard rock (e.g., volcanic, igneous and sedimentary), although softer geologic materials such as salt domes and thick clay sequences are also under consideration. Low-permeability rock masses have little or negligible matrix porosity and permeability, with connected networks of discontinuous fractures imparting secondary porosity and permeability. Second, predicting radionuclide transport in fractured media is a formidable challenge as the spatially-discontinuous nature of fracture networks, along with high degrees of heterogeneity within fracture properties, lead to highly anisotropic flow systems with complex patterns of fluid flow and subsequent radionuclide migration (de Dreuzy et al., 2001; Klimczak et al., 2010; Neuman, 2005; Reeves et al., 2008a;b;c; Schwartz et al., 1983; Smith & Schwartz, 1984). Detailed radionuclide transport predictions are typically reliant on numerical simulations that incorporate site-specific fracture data (e.g., Arnold et al., 2003; Bodvarsson et al., 2003; Cvetkovic et al., 2004; Pohll et al., 1999; Pohlmann et al., 2002; 2004; Reeves et al., 2010; Robinson et al., 2003; Smith et al., 2001), though some analytical techniques for first-cut approximations have been developed (Reeves et al., 2008b; Zhang et al., 2010).

This Chapter is not designed to provide a full treatise on the characterization of rock masses for development of complex numerical models used to predict radionuclide transport. Rather, we present an alternative approach where site-specific fracture network properties can be used to infer flow and transport characteristics of fracture networks by expanding the framework proposed by Reeves et al. (2008c). This framework can be used to qualitatively evaluate the suitability of candidate rock masses intended for the disposal of high-level radioactive waste based on fracture statistics. In stark contrast to current evaluation approaches that rely on costly field investigations to supply data to numerical models for radionuclide transport predictions, this type of evaluation promotes both time and economic savings by screening candidate fractured rock masses according to relatively simple criteria obtained from fracture characterization efforts.

2. Fractured rock characterization

Fractures are spatially discontinuous features that exhibit strong variability in geometrical and hydraulic properties. This variability is a result of the complex interplay between current and past stress fields, rock mechanical properties (i.e, Young's modulus, Poisson's ratio), mechanical fracture interaction and distributions of flaws or weakness in a rock mass. Fractured rock masses are typically characterized during field campaigns that measure fracture attributes from a number of sources including boreholes, rock outcrops, road cuts and tunnel complexes. Seismic techniques can also be used to image fault structures in the subsurface. Hydraulic properties of fractured media can be either inferred from fracture aperture or hydraulic tests performed on boreholes.

Full characterization of fractured rock masses is not possible since known fracture locations and their attributes consist of an extremely small sample of the overall fracture network, i.e., any fracture characterization effort grossly undersamples a field site due to limited accessibility to the fractures themselves. Fracture data, however, can be used to generate representative, site-specific fracture networks through the derivation of probabilistic descriptions of fracture location, orientation, spacing, length, aperture, hydraulic conductivity/transmissivity and values of network density (Figure 1). With the exception of adsorption, these are the statistical properties that form the basis of the geologic repository screening framework. Statistical analysis of fracture attributes will be extensively covered in this section.

2.1 Orientation

Fracture networks will typically have two or more fracture sets characterized by fracture orientation (e.g., Barton, 1995; Bonnet et al., 2001; Bour & Davy, 1999; Pohlmann et al., 2004; Reeves et al., 2010). The presence of at least two intersecting sets of fractures reflects the physics of rock fracture propagation where two sets of fractures can arise from a single stress field (Jaeger et al., 2007; Twiss & Moores, 2007). Unless fractures are very long such as regional-scale faults, it is important from a flow perspective to have at least two intersecting fracture sets to promote connectivity across a rock mass.

The orientation of fracture planes is denoted by strike and dip convention. Analysis of fracture orientation begins with projecting the poles to fracture planes onto a stereonet and using contours of pole density to identify fracture sets (Figure 2). Upon identification of fracture sets, mean orientation and the variability of fracture poles for each fracture set can be determined.

The distribution of fracture orientation is usually modeled using a Fisher distribution (Fisher, 1953):

$$f(x) = \frac{\kappa \cdot \sin x \cdot e^{\kappa \cdot \cos x}}{e^{\kappa} - e^{-\kappa}} \tag{1}$$

where the divergence, x (degrees), from a mean orientation vector is symmetrically distributed $(-\frac{\pi}{2} \leq x \leq \frac{\pi}{2})$ according to a constant dispersion parameter, κ. The Fisher distribution is a special case of the Von Mises distribution, and is similar to a normal distribution for spherical data (Mardia & Jupp, 2000). The extent to which individual fractures cluster around the mean orientation is described by κ where higher values of κ describe higher degrees of clustering. It is our experience that values of κ are commonly in the range of $10 \leq \kappa \leq 50$ for natural fracture networks. Stochastic simulation of Fisher random deviates in the discrete fracture networks following this section is based on the method proposed by Wood (1994).

The Bingham distribution provides an alternative to the Fisher distribution for cases in which fracture strike and dip are asymmetrically clustered around mean fracture orientations (Bingham, 1964):

$$f(x) = \frac{exp\left[\kappa_1(M_1 x)^2 + \kappa_2(M_2 x)^2 + \kappa_3(M_3 x)^2\right]}{4\pi \cdot F(1/2; 3/2; E)} \tag{2}$$

where κ_1, κ_2 and κ_3 are dispersion coefficients that satisfy the condition: $\kappa_1 \leq \kappa_2 \leq \kappa_3 = 0$, M_1, M_2 and M_3 are the column vectors of matrix $\mathbf{M}$, $\mathbf{E}$ is the eigenvector matrix, and $F(1/2; 3/2; E)$ is a hypergeometric function of the matrix argument. The probability distribution function described by (2) can especially occur for faults that exhibit a greater range in deviations in strike than dip. A shortcoming of the Bingham distribution, however, is that it is not

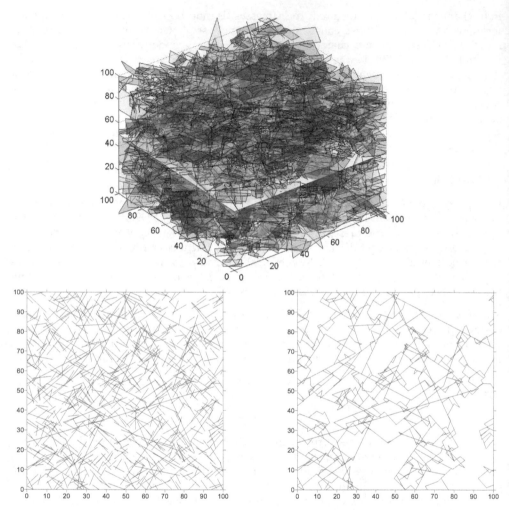

Fig. 1. Illustration showing the correspondence between two- and three-dimensional fracture networks. The three-dimensional network (top) is generated according to two fracture sets with significant variability about mean fracture orientations, a power-law length distribution exponent of $a = 2.0$ and a relatively sparse density. The two-dimensional network at the bottom left is computed by projecting all fractures onto the yellow horizontal slice located in the center of the three-dimensional DFN. The two-dimensional network on the bottom right is the result of identifying the hydraulic backbone by eliminating all dead-end fracture segments and non-connected clusters. Once the hydraulic backbone is identified, flow and particle transport can then be computed for the network.

mathematically possible to use (2) for the stochastic generation of asymmetric deviates. The authors are currently developing a method to simulate asymmetric deviations from mean fracture set orientations.

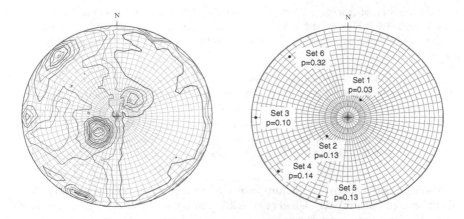

Fig. 2. Stereonet plots of poles to fracture planes with contour plots of all poles (left) and identified fracture sets along with prior probability (right). From Reeves et al. (2010).

2.2 Spacing

Fracture spacing refers to the linear distance between fractures. This distance also provides a length scale for unfractured matrix blocks. Fracture spacings from a data set require a correction (Terzaghi, 1965):

$$D = D' sin(\alpha) \tag{3}$$

to convert the apparent spacing D' measured along a transect to true fracture spacing D. Values of α denote the angle of the transect relative to the mean fracture orientation or a pre-determined reference direction (Figure 3). Apparent spacing is equal to true spacing if the transect is perpendicular to the mean fracture orientation or reference direction. If $\alpha = 90°$ the Terzaghi correction factor $f = sin(\alpha)$ reduces to 1.

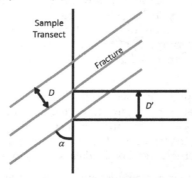

Fig. 3. Illustration showing how the Tergazhi correction accounts for the bias between apparent spacing D' and true spacing D based on the orientation of the sample transect in relation to mean fracture orientation. After http://www.rocscience.com/downloads/dips/WebHelp/dips/Terzaghi_Weighting.htm.

Once the spacing between fractures is corrected, values of fracture frequency and average fracture spacing can be computed. Fracture frequency [units of inverse length, L^{-1}] is defined as the total number of fractures along the distance of a transect. Average fracture spacing [units of length, L] is simply the inverse of fracture frequency, and defines the size of the unfractured matrix block. This metric and its influence on radionuclide retention will be discussed in a later section.

Spacing in natural fracture networks is most commonly an exponentially distributed random variable. This can be tested by plotting the inverse empirical cumulative distribution function of fracture spacing, also known as a survival function (Ross, 1985). If spacing is exponential, the probability decay of the tail of the empirical fracture spacing distribution will exhibit a straight line on a semi-log plot. A Poisson point process defined by independent and identically distributed uniform deviates provides easy generation of exponential spacing (Ross, 1985).

Other possible distributions of fracture spacing include uniform (Rives et al., 1992) and fractal clustering (Barton, 1995; Darcel et al., 2003), both of which are considered extreme end members. Uniform spacing may occur in thin geologic layers which restrict fracture growth in the vertical direction and promote long horizontal fracture growth with nearly constant spacing (Rives et al., 1992). Exact causes of fractal clustering are less known and may be related to the role of mechanical fracture interaction during propagation that likely controls fracture length and spacing (Ackermann & Schlische, 1997; Darcel et al., 2003; Olson, 1993; Segall & Pollard, 1983). Networks with fractal clustering can be generated via a multiplicative cascade process (Mandelbrot, 1974; Schertzer & Lovejoy, 1987).

2.3 Length

Fracture length denotes the horizontal trace length of a fracture. There is a consensus in recent literature that fracture lengths above a lower length cutoff, l_{min}, are power-law:

$$P(L > l) = Cl^{-a} \tag{4}$$

with a power law exponent, a, that ranges between 1 and 3 in natural fracture networks (Bonnet et al., 2001; Bour & Davy, 1997; 1999; Renshaw, 1999). C is a constant based on l_{min} and a. Though lognormal distributions of fracture length have been reported in the literature, they are a result of improper sampling of the largest fractures within a sampling window. Lognormal distributions easily arise in data sets with power-law tails if the largest values are censored.

Determination of the distribution of fracture length is similar to that of fracture spacing and involves the analysis of an inverse empirical cumulative distribution function. Fracture lengths that are power-law will exhibit linear trends on a log-log plot for the tail of the distribution. In this example, the tail of the distribution refers to the greatest 5-10% of length values. The slope of the power-law trend of the data is equal to $-a$.

Truncations can frequently occur in fracture length data due to constraints imposed by the finite scale of the sampling window. For example in Reeves et al. (2010), the longest fracture measured in a tunnel drift was parallel to the drift and was approximately two-thirds of the total drift length. Instead of choosing between a traditional power-law or lognormal distribution, an upper truncated Pareto (power law) model (Aban et al., 2006):

$$P(L > l) = \frac{\gamma^a(l^{-a} - v^{-a})}{1 - (\frac{\gamma}{v})^a} \tag{5}$$

was used to compute the power-law trend in the data, where $L_{(1)}, L_{(2)}, \ldots, L_{(n)}$ are fracture lengths in descending order and $L_{(1)}$ and $L_{(n)}$ represent the largest and smallest fracture lengths, γ and ν are lower and upper fracture length cutoff values, and a describes the tail of the distribution. Truncated power law models like (5) can also be useful for imposing an upper length scale to the generation of stochastic networks at the regional scale. Lacking evidence of domain-spanning faults (with the exception of bounding faults of the stock itself) for a 5 km wide granitic stock, Reeves et al. (2010) assigned an upper limit of 1 km in the stochastic generation of fault networks.

2.4 Hydraulic conductivity

Boreholes are commonly used to characterize fractured media. Borehole geophysics can provide useful information about fractures within the rock including fracture frequency, orientation, aperture and mineral infilling. Fracture aperture, defined as the width of the void space normal to fracture walls, can be used to infer hydraulic properties of fractures. Fracture apertures at land surface have low confining stresses that are not representative of subsurface confining stresses within a rock mass. We therefore recommend that fracture aperture values used to compute flow are measured in boreholes where in-situ stress is preserved.

The cubic law, a solution to the Navier-Stokes equation for laminar, incompressible flow between two parallel plates, describes a general relationship between fluid flow and fracture aperture (Snow, 1965):

$$Q = -\frac{\rho g}{12\mu} b^3 \nabla h \qquad (6)$$

where fluid discharge per unit width, Q [L^2/t], is proportional to the cube of the hydraulic aperture, b. Similar to Darcy's Law, the cubic law (6) assigns discharge through a fracture as a linear function of the hydraulic gradient, ∇h. The relationship between hydraulic aperture and transmissivity (T), hydraulic conductivity (K) and permeability (k) is described by: $T = \frac{\rho g}{12\mu} b^3$, $K = \frac{\rho g}{12\mu} b^2$ and $k = \frac{b^2}{12}$, respectively. Fluid-specific properties density, ρ, and viscosity, μ, allow for conversions between permeability and hydraulic conductivity/transmissivity. As a note of caution, the relationship between mechanical aperture, the physical distance between fracture walls, and hydraulic aperture, the equivalent aperture for a given flow rate, is unclear. As a general rule, hydraulic aperture is typically smaller than mechanical aperture (Chen et al., 2000; Cook et al., 1990; Renshaw, 1995; Zimmermann & Bodvarsson, 1996). Discrepancies between mechanical aperture and hydraulic aperture are attributed to surface roughness, flow path tortuosity, and stress normal to the fracture. Though empirical correction factors have been used to correlate mechanical and fracture apertures (Bandis et al., 1985; Cook et al., 1990; Renshaw, 1995), no method is reliable for a wide range of aperture values.

Hydraulic testing of boreholes yields reliable estimates of fracture T and K. While there are many different hydraulic testing techniques, the isolation of specific intervals during testing with the use of dual-packer systems provides the best data to characterize the distribution of transmissivity/hydraulic conductivity. These tests yield flow rate information for applied fluid pressures, which also allows for the inverse computation of hydraulic aperture using (6). These aperture values, in addition to T and K estimates, are useful for parameterizing flow and transport models.

Studies in highly characterized rock masses have shown that fracture K is extremely heterogeneous and may encompass 5 to 8 orders of magnitude (Andersson et al., 2002a;b;

Guimerá & Carrera, 2000; Paillet, 1998). Often the distribution of K (and T) is thought to be lognormal:

$$p(x) = \frac{1}{x\sqrt{2\pi\sigma^2}}exp\left[\frac{-(\log x - \mu)^2}{2\sigma^2}\right] \tag{7}$$

where x is the mean and σ is the standard deviation. Values of $log(\sigma_K)$ are typically around 1 for fractured media (Andersson et al., 2002a;b; Stigsson et al., 2001). However, other studies suggest power law distributions (Gustafson & Fransson, 2005; Kozubowski et al., 2008) and that these lognormal distributions, similar to length, could be caused by censoring flow data possibly due to instrument limitations. Additionally, flow through rough-walled fractures can be non-Darcian (Cardenas et al., 2007; Qian et al., 2011; Quinn et al., 2011). This may further complicate the estimation of hydraulic conductivity in field hydraulic tests as flow is no longer linearly proportional to a pressure gradient as described by (6).

2.5 Density

Fracture networks consist of two-dimensional planes embedded within a rock matrix (Figure 1). The lack of access to the total rock volume makes it impossible to directly measure the three-dimensional fracture density of a rock mass. Instead, three-dimensional density for discrete fracture networks is estimated from density measurements of lower dimensions, i.e., one-dimensional fracture frequency from boreholes and/or tunnel drifts or two-dimensional fracture density from outcrops and fracture trace maps. Definitions of fracture density according to dimension are: one-dimensional density (also known as fracture frequency), ρ_{1D} [L^{-1}], is expressed as the ratio of total number of fractures, f_i, to transect length, L: $\rho_{1D} = L^{-1}\sum_{i=1}^{n} f_i$; two-dimensional fracture density, ρ_{2D} [L/L^2], is expressed as the ratio of the sum of fracture lengths, l, to area, A: $\rho_{2D} = A^{-1}\sum_{i=1}^{n} l_i$; and three-dimensional fracture density, ρ_{3D} [L^2/L^3], is expressed as the ratio of the sum of fracture plane area, A_i, to rock volume, V: $\rho_{3D} = V^{-1}\sum_{i=1}^{n} A_i$.

Numerical techniques can be used to upscale one-dimensional fracture frequency [L^{-1}] estimates to a three-dimensional spatial density [L^2/L^3] (Holmén & Outters, 2002; Munier, 2004). For example, one-dimensional transects can be used to upscale two-dimensional networks by adding fractures until the one-dimensional transect density is satisfied along several transects placed along the two-dimensional network. Three-dimensional networks can be generated in a similar fashion by either generating fractures until the frequency along one-dimensional boreholes is satisfied or by projecting fractures onto sampling planes (e.g., Figure 1) until a two-dimensional density criterion is satisfied.

Fracture density is highly dependent on the distribution of fracture lengths in a model domain where the density at the percolation threshold increases with increasing values of a (Darcel et al., 2003; Reeves et al., 2008b; Renshaw, 1999). This will become apparent in the fracture network examples in the next section.

3. Flow and advective transport properties of fracture networks

A central theme of this Chapter is that flow and radionuclide transport characteristics of fractured media can be inferred from fracture statistics. The previous section discussed in detail how fractured rock masses can be analyzed according to statistics of fracture orientation, spacing, length, hydraulic conductivity and values of fracture density. This section contains detailed explanations on how these attributes can provide *a priori* insight into flow and transport properties of a fractured rock mass. We begin by defining network structure, and then discuss how the structure relates to transport characteristics such as shape of breakthrough curves, potential for early arrivals, and variability of individual breakthroughs about the ensemble. Simulations of flow and transport in two-dimensional discrete fracture networks (DFN) with physically realistic parameters are used to illustrate specific concepts.

3.1 Network structure and transport

The structure of natural fracture networks is the end result of the complex interplay between stress fields and their anisotropy, mechanical properties of the rock, mechanical fracture interaction and distribution of initial flows in a rock mass. The reliance on probabilistic descriptions of fracture attributes reflects our lack of ability to accurately construct fracture networks based on mapping studies alone. The limited accessability to the network leaves an incomplete understanding of the patterns of fracturing within a rock mass that can often be improved through visual inspection of representative networks generated according to site-specific statistics.

A total of three different network types are generated from two fracture sets with power law distribution of lengths with exponent values in the range $1.0 \leq a \leq 3.0$, fracture density, $1.0 \leq \rho_{2D} \leq 2.0$ m/m^2, orientations of $\pm 45°$ with variability described by $\kappa = 20$ and a lognormal hydraulic conductivity distribution with $\log(\sigma_K) = 1$ (Figures 4–5). Once a network is generated, the hydraulic backbone is identified by eliminating dead-end segments and isolated clusters. This is accomplished in our model through both geometric and flow techniques. The hydraulic backbone represents the interconnected subset of a fracture network that is responsible for conducting all flow and transport across a domain. Hence, analysis of backbone characteristics can provide insight into these processes.

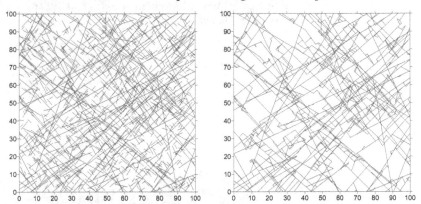

Fig. 4. Discrete fracture network realization (left) and hydraulic backbone (right) generated from two fracture sets with power distribution of lengths with exponent $a = 1.0$, $\rho_{2D} = 1.0$ m/m^2, $l_{min} = 2.0$ m and orientations at $\pm 45°$ with variability described by $\kappa = 20$. Note that the hydraulic backbone is dominated by long fractures.

The generated networks in this study do not explore the full parameter space for fractured media. However, the wide range of fracture length exponents provides sufficient variability and produces three distinct types of hydraulic backbones. Networks generated with $a = 1.0$

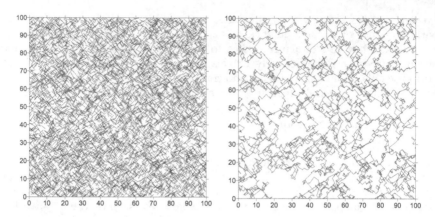

Fig. 5. Discrete fracture network realization (left) and hydraulic backbone (right) generated from two fracture sets with power distribution of lengths with exponent $a = 3.0$, $\rho_{2D} = 2.0$ m/m^2, $l_{min} = 2.0$ m and orientations at $\pm 45°$ with variability described by $\kappa = 20$. Note that the hydraulic backbone is dominated by short fractures.

produce backbone structures dominated by long fractures (Figure 4), and networks with $a = 3.0$ produce backbone structures dominated by short fractures (Figure 5). Backbones with a mixture of short and long fractures are produced for networks generated with $a = 2.0$ (Figure 6). Another feature of these networks is that density of the network increases from $\rho_{2D}=1.0$ m/m^2 to $\rho_{2D}=2.0$ m/m^2 as the value of a increases from 1.0 to 3.0. This increase in density is necessary to maintain a percolating backbone. For example, the density values assigned to $a = 1.0$ and $a = 2.0$ ($\rho_{2D} =1.0$ and 1.5 m/m^2, respectively) result in a non-percolating networks if used with $a = 3.0$. Conversely, networks generated with $a = 1.0$ and $\rho_{2D} = 2.0$ m/m^2 (assigned to $a = 3.0$) produces unrealistically dense networks.

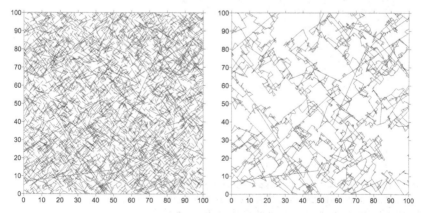

Fig. 6. Discrete fracture network realization (left) and hydraulic backbone (right) generated from two fracture sets with power distribution of lengths with exponent $a = 2.0$, $\rho_{2D} = 1.5$ m/m^2, $l_{min} = 2.0$ m and orientations at $\pm 45°$ with variability described by $\kappa = 20$. Note that the hydraulic backbone is a mix of short and long fractures.

Computation of flow in two-dimensional discrete fracture networks involves solving for hydraulic head at all internal nodes (intersection point of two or more fractures) inside the domain according to Darcy's law (de Dreuzy & Ehrel, 2003; Klimczak et al., 2010; Priest, 1993). In the simulations presented, boundary conditions are applied to all nodes on the domain to induce flow from top to bottom according to a linear hydraulic gradient of 0.01. Fluid flow through the backbone is then solved iteratively at each node via a biconjugate gradient method (Figure 7). A total of 500,000 conservative (non-sorbing) radionuclide particles are then introduced at the top of the domain, traced through the network, and recorded as they exit the model domain. The velocity along each fracture segment is constant, and radionuclide particles are moved by advection from node to node. Radionuclide particles change direction at nodes proportional to the flux of the down gradient fracture segments. This process is repeated to produce 50 statistically equivalent networks and resultant breakthrough curves for each of the three network types.

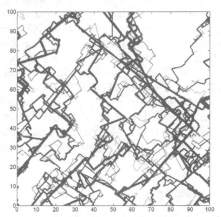

Fig. 7. Discrete fracture network generated according to parameters specified in the caption to Figure 6 and a lognormal distribution of hydraulic conductivity with $\log(\sigma_K) = 1.0$. Line thickness is proportional to flow through each segment. Note the large contrasts in flow, and hence transport velocities, between different segments of the hydraulic backbone. Particles sample segments of different velocity at intersections where particles are routed proportional to segment flux.

The radionuclide breakthroughs reflect the characteristics of the fracture networks (Figure 8). First, fracture length defines the correlation structure of the network. Lower values of a place more probability weight on extreme values of fracture length, and consequently these networks consist of very long fractures that often span the model domain. Radionuclides traveling through these networks only change direction at fracture intersections, and the combination of long fractures and sparse density promotes fast transport. This can be observed in Figure (8) where mean ensemble breakthrough occurs at 1.5 years for a transport distance of 100 meters. Networks with $a = 2.0$ assign less probability weight on extreme values and produce backbones with a mixture of short and moderately long fracture lengths. This backbone structure requires radionuclides to move through a mixture of short and long fractures and sample a wider variety of velocities in the process of migrating across the model domain. Consequently, mean ensemble breakthrough for networks with $a = 2.0$ occurs at 11 years, which is approximately 7 times slower than for $a = 1.0$ (Figure 8). The slowest transport occurs for networks with $a = 3.0$ which consist of backbones dominated by short fracture segments. The short segments and higher density causes radionuclide particles to sample a very large number of fracture segments and velocities in the process of migrating across the domain. The mean breakthrough for $a = 3.0$ networks occurs at 26 years, which is approximately 17 and 2.4 times slower than the $a = 1.0$ and $a = 2.0$ networks (Figure 8).

Second, the breakthrough curves for individual realizations all show a high degree of variability (Figure 8). In fact, this variability is the reason why breakthroughs are presented in a cumulative form, as the shape of the breakthroughs can range from sharp peaks (more prevalent for networks with $a = 1.0$) to more broad "Gaussian-like" plumes (more prevalent for networks with $a = 3.0$). Networks with $a = 2.0$ represent an intermediate case where breakthroughs for individual realizations can encompass both scenarios. All of the networks exhibit heavy late-time tails which are briefly discussed in the following section. The variability in fracture networks is quantified by comparing deviations of individual realizations from the ensemble. Specifically, the standard deviation of breakthroughs is computed for cumulate probabilities ranging from 0.05 to 0.40 in intervals of 0.05 (Figure 9). This metric is then normalized by the ensemble mean for comparison to the other network types, and termed realization variability. Trends of realization variability are highest for $a = 1.0$ and lowest for $a = 3.0$. The largest realization variability values occur for early breakthroughs (0.05 to 0.10 of total mass) for $a = 1.0$ and $a = 2.0$. This can be attributed to the fact that these networks, especially $a = 1$, produce early arrivals that vary greatly in time. Reeves et al. (2008c) studied deviations in spatial distribution of individual solute plumes from the ensemble, and their results support our findings that networks with $a = 1.0$ are inherently unpredictable, and predictability in transport increases as mean fracture length decreases.

4. Radionuclide retention mechanisms

The previous section illustrates how statistics of fractured massses can be used to infer advective characteristics of radionuclide transport. In this section, we address the retentive properties of fractured rock masses which include adsorption of radionuclides onto fracture walls and within matrix blocks, and diffusional mass exchange of radionuclides between fractures and low-velocity zones. Adsorption is a chemical process that is related to the mineralogy of the host rock, and cannot be inferred from fracture network statistics. One must also consider that several radionuclides, such as ^3H, ^{14}C, ^{36}Cl, ^{39}Ar and ^{85}Kr, are conservative and not subject to sorption processes. The dominant retention mechanism for conservative radionuclides is diffusional mass exchange. The retention characteristics of diffusional mass exchange of radionuclides between fractures and matrix blocks can be inferred from fracture network properties and will receive the majority of attention in this section. A short discussion on sorption is included for completeness in the context of the screening framework. A network-scale retention mechanism arising from velocity contrasts among conductive fracture segments and is also introduced for additional consideration.

4.1 Diffusional mass exchange

Molecular diffusion promotes the exchange of radionuclides between active fracture flow paths and low-velocity zones such as stagnant zones within fractures, fault gauge and unfractured matrix blocks. Among these low-velocity regions, the influence of diffusional mass exchange of radionuclides in unfractured matrix blocks is much greater than diffusional

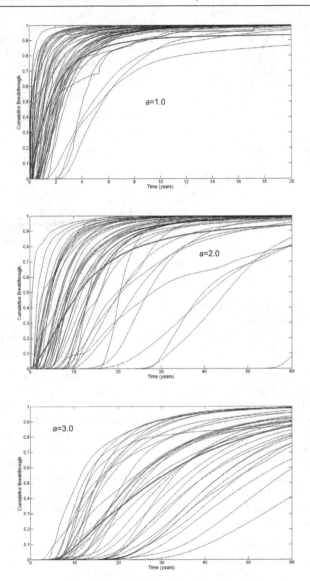

Fig. 8. Normalized conservative particle breakthroughs for 100 m scale transport in two-dimensional networks with power-law exponents $a = 1.0$ (top), $a = 2.0$ (middle) and $a = 3.0$ (bottom) for 50 transport realizations. Individual realizations are represented as thin black lines with the blue curve representing the ensemble. Note that mean arrival times increase with increases in power-law exponent, where $a = 1.0$, $a = 2.0$ and $a = 3.0$ lead to mean arrival times of 1.5, 11 and 26 years, respectively. Deviations in breakthroughs also increase with increasing values of a. The smooth breakthroughs in Figure 8 reflects the high number of particles.

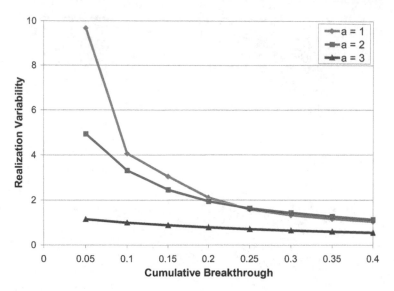

Fig. 9. Plot of realization variability in particle breakthroughs about ensemble breakthough for networks generated according to power law length distribution exponents of $a = 1.0$, $a = 2.0$, and $a = 3.0$. Note that realization variability decreases as the value of the power law length exponent decreases.

exchange within fracture stagnant zones and fault gauge (Andersson et al., 2002a;b; Grisak & Pickens, 1980; Neretnieks, 1980; Sudicky & Frind, 1982; Winberg et al., 2000). The reason for this is simple: the time a radionuclide molecule spends trapped in an immobile zone is proportional to the size of that zone, and the scale of matrix blocks is typically on the order of centimeters to meters while within fracture stagnant zones and fault gauge are on the order of micrometers to millimeters. Diffusion into matrix blocks was found to be a major retention mechanism at the time scales of the Tracer Retention Understanding Experiments conducted at the Äspö underground laboratory at both single fracture (Winberg et al., 2000) and block scales (Andersson et al., 2002a;b). For these in-situ experiments, the prevalence of molecular diffusion as a retention mechanism is exhibited by late-time breakthroughs with slopes equal to -3/2 for both conservative and non-conservative radionuclides.

Fracture spacing is a network parameter that denotes unfractured matrix block size and provides a length scale for molecular diffusion. The influence of fracture spacing on contaminant breakthroughs is illustrated in Figure 10 for ^3H. The figure is generated from an exact analytical solution to conservative transport through a series of 40 m parallel-plate fractures with 110 μm apertures subject to dual-domain mass transfer through fractures and matrix blocks (Sudicky & Frind, 1982). The mean age of ^3H breakthroughs versus fracture spacing (2B) is shown in Figure 11. Figures 10 and 11 show that increases in mean breakthrough time for conservative radionuclides are highly correlated to fracture spacing with a power-law (nearly linear) trend. These results indicate that the retention potential for a candidate repository rock mass can be inferred from fracture spacing, and that larger average spacing leads to greater retention due to diffusional mass transfer between fractures and matrix.

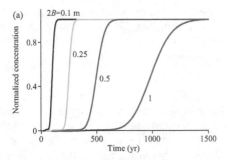
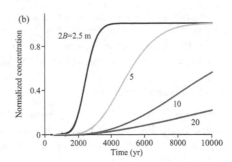

Fig. 10. Normalized ^{3}H breakthroughs for a series of 40 m parallel-plate fractures with 110 μm apertures ($K = 1.0 \times 10^{-9}$m/s) for a variety of matrix block sizes (2B). A diffusion coefficient for tritiated water at 25°C ($D = 2.4 \times 10^{-08}$ m^2/s) with a matrix tortuosity and porosity of 0.25 and 0.1, respectively, were used to generate the breakthroughs from analytical solutions (Sudicky & Frind, 1982). Note the dramatic affect of fracture spacing on radionuclide breakthrough due to diffusional mass exchange between fractures and matrix blocks.

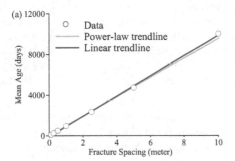
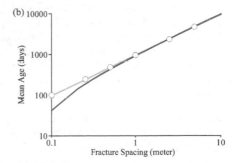

Fig. 11. Mean age (A) of ^{3}H exiting the fracture versus fracture spacing (2B) from Figure 10 with best-fit linear (red) and power-law (green) trendlines. The slightly non-linear trend in mean age versus spacing is more apparent in log-log axes (b). The best-fit power-law trendline is: $A = 971(2B)^{0.9934}$.

Although Figures 10 and 11 are helpful in illustrating and partially quantifying the influence of fracture spacing on radionuclide retention by diffusional mass transfer, it is important to acknowledge that fractured media is inherently much more complex and many simplifying assumptions were made in this analysis. First, cystalline rocks typically have very low porosities and this void space may become disconnected further into the matrix block, giving rise to the concept of maximum penetration depth which describes a cutoff to the distance radionuclides can diffuse into the matrix block (e.g., Neretnieks, 1980). Thus, for crystalline rock masses, average fracture spacing would have to reflect penetration depth which is usually unknown. Second, fracture networks are discontinuous in nature and exhibit power-law fracture lengths which promote a broad distribution of matrix block sizes. For these networks, a "characteristic" or average fracture spacing does not exist and the retention of radionuclides in matrix blocks of different sizes can result in a continuum of exchange rates that are both faster and slower than the what would be predicted using the arithmetically computed average block size. Third, weathering processes at the interface of a fracture and the host rock can create higher porosity rim zones that enhance diffusion (Cvetkovic, 2010). Despite these complicating factors, it can be generally stated that rock masses with greater average fracture spacing have greater potential for radionuclide retention.

4.2 Radionuclide adsorption

There may be cases where potential repositories have similar fracture characteristics, the most important being the distribution of fracture length, network density and average fracture spacing, and the sorption capacity of the host rock may serve as a secondary screening criterion. Adsorption of radionuclides is a reversible chemical process that can include both cation exchange and surface complexation. It can often be difficult to distinguish in-situ between retention processes as molecular diffusion plays a role in introducing sorbing radionuclides to adsorptive sites within fractured rock environments. The use of multiple tracers, including conservative and non-conservative with varying sorbing rates, can be used to ascertain the relative contributions of retention from diffusional mass transfer and adsorption (e.g., Andersson et al., 2002b).

Adsorption in fractured rock can occur along fracture walls and within fault gauge and matrix blocks. Sorption along fracture walls is a surface process rather than a volumetric process, like sorption within matrix blocks. Similar to the discussion on molecular diffusion, the TRUE-1 and TRUE Block Scale tests conducted at Äspö Hard Rock Laboratory indicate that sorption in the unfractured rock matrix is a dominant process over adsorption onto fracture walls and fault gauge (Andersson et al., 2002a;b). This is because matrix blocks have a greater surface area and length scale, and hence, a greater number of available sorption sites than fracture walls and fault gauge.

Sorption is typically characterized using equilibrium isotherm models parameterized via laboratory batch experiments containing soil, sediment or crushed rock from a site of interest (e.g., Langmuir, 1997; Zheng & Bennet, 2002). This approach to modeling sorption assumes that rates of sorption/desorption between radionuclides and the sorptive medium are orders of magnitude more rapid than the advective groundwater velocity. However, recent studies have shown that desorption rates of U and [237]Np may yield a 3-4 order of magnitude range of desorption rate constants despite uniform flow fields approximating equilibirum conditions (Arnold et al., 2011; Dean, 2010). We recommend from a cost savings perspective that candidate rock masses be screened on the basis of sorption coefficients determined from laboratory batch experiments. However, the findings of large desorption rate constants questions the usefulness of characterization techniques relying on equilibrium assumptions for use in radionuclide transport models.

4.3 Network-scale retention

Numerical investigations have shown that velocity contrasts within fracture segments on the hydraulic backbone can lead to solute retention within the backbone itself (Berkowitz and Scher, 1997; Painter & Cvetkovic, 2002; Reeves et al., 2008c). Painter & Cvetkovic (2002) and Reeves et al. (2008c) found that distributions of inverse velocity, $1/v$, a surrogate for retention, exhibited power law decay trends at late times. Values of γ measured from the tail of the distribution of $1/v$ versus time were in the range $1.1 \leq \gamma \leq 1.8$ for Painter & Cvetkovic (2002) and approximately $\gamma=1$ for Reeves et al. (2008c). Analysis of the survival function of ensemble breakthrough versus time for the $a = 1.0$ networks in this study yielded a value of $\gamma = 1.5$ (Figure 12). Values of γ for $a = 2.0$ and $a = 3.0$ networks were not computed due to incomplete breakthroughs that leave the tail of the distribution undefined. None of these studies found a correlation between distributions of fracture length and γ; however, all of these studies used transmissivity distributions encompassing several orders of magnitude. It is entirely possible that this type of retention results from the interaction of power law distributions of trace length and distributions of transmissivity that vary by several orders of magnitude, such as lognormal or power law. More work in this area is needed to provide further insight into this retention process in the context of fracture statistics.

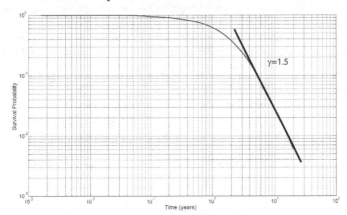

Fig. 12. Survival function (inverse empirical cumulative distribution function) of particle breakthrough versus time for the $a = 1.0$ networks generated in this study (blue) along with best-fit tail trendline (black). The trendline scales according to $-1 - \gamma$ where a slope of -2.5 corresponds to $\gamma=1.5$.

5. Geologic repository screening framework

The geologic barrier relies on the intrinsic ability of the host rock to limit the release of radionuclides to the biosphere. The discussion up to this point has included both advective and retentive characteristics of fractured media. Desirable qualities for a potential repository rock mass from a radionuclide transport perspective include: low potential for early arrivals, longer bulk breakthroughs, and low variability in predicted results. Simulated particle breakthroughs show a clear trend that rock masses with networks dominated by short fractures exhibit advective transport characteristics that are most desirable for geologic repositories. These networks have the lowest potential for early arrivals, breakthroughs occur at much longer timescales than networks consisting of longer fractures, and the variability of individual realizations about the ensemble are lower than for networks generated using $a = 2.0$ and $a = 3.0$. These conclusions are supported by the study of Reeves et al. (2008c).

Retention of radionuclides via molecular diffusion favors networks with large fracture spacings, as retention time for conservative radionuclides in a low-velocity matrix block is proportional to block size. This suggests that the sparsest fracture networks are the most desirable. Given the link between fracture density and the distribution of trace length, networks dominated by long fractures (e.g., $a < 2.0$) would be the optimum choice.

However, these networks produce breakthroughs that exhibit the earliest arrivals and bulk transport times, and have the greatest variability. It is our opinion that advective transport characteristics should be weighted higher than retention since slower advective transport rates through a network decreases the propensity for fast radionuclide transport and early arrivals. With these network types, retention processes such as molecular diffusion and adsorption will further retard the relatively slow migration rates of radionuclides. Note that molecular diffusion has the ability to significantly retard contaminant migration even for relatively small matrix block sizes (Figures 10 and 11). Networks with identical length distributions can be distinguished from another on the basis of average fracture spacing, a secondary criterion. The rock mass with larger values of average spacing (assuming similar host rock porosity) will have a greater retention potential via molecular diffusion. Sorption characteristics of a rock mass may serve as a tertiary criterion, and this consideration must take into the account the waste itself and the prevalence of non-conservative radionuclides.

This screening framework is intended to be cost effective as flow and transport characteristics of fractured rock masses are inferred solely on the basis of fracture statistics. The selection of an ideal geologic repository for high-level radioactive waste disposal is a complex endeavor that must include considerations of geologic stability (e.g., lack of recent seismic events, volcanic activity, high geothermal gradients), hydrology, rock mechanical strength, geochemistry and proximity to large population centers and natural resources (e.g., National Research Council, 1978). It is assumed that these factors will be considered at some point in the selection process.

6. References

Aban, I.B.; Meerschaert, M.M. & Panorska, A.K. (2006). Parameter estimation methods for the truncated Pareto distribution, *J. Amer. Stat. Assoc.*, Vol. 101, 270–277.

Ackermann, R.V. & Schlische, R.W. (1997). Anticlustering of small normal faults around larger faults, *Geol.*, Vol. 25, No. 12, 1127–1130.

Andersson, P.; Byegärd, J.; Dershowitz, B.; Doe, T.; Hermanson, J.; Meier, P.; Tullborg, E.-L.; & Winberg, A. (2002a). *Final Report on the TRUE Block Scale Project: 1. Characterization and Model Development*, Swedish Nuclear Fuel and Waste Management Company (SKB), Technical Report TR-02-13, Stockholm, Sweden.

Andersson, P.; Byegärd, J.; & Winberg, A. (2002b). *Final Report on the TRUE Block Scale Project: 2. Tracer Tests in the Block Scale*, Technical Report TR-02-14, Swedish Nuclear Fuel and Waste Management Co. (SKB), Stockholm, Sweden.

Arnold, B.W.; Kuzio, S.P.; & Robinson, B.A. (2003). Radionuclide transport simulation and uncertainty analysis with the saturated-zone site-scale model at Yucca Moutain, Nevada, *J. Contam. Hydrol.*, Vol 62–63, doi:10.1016/S0169-7722(02)00158-4.

Arnold, B.; Reimus P.; James, S. (2011). *Flow and Transport in Saturated Media: FY2011 Status Report*, prepared fro the U.S> Department of Energy Used Fuel Disposition Campaign, FCRD-USED-2011-0003111, Sandia National Laboratories and Los Alamos National Laboratory.

Bandis, S.C.; Makurat A. & Vik, G. (1985). Predicted and measured hydraulic conductivity of rock joints, *Proceedings of the International Symposium on Fundamentals of Rock Joints*, Björdkliden, Norway, Septemeber 15-20.

Barton, C.C. (1995). Fractal analysis of scaling and spatial clustering of fractures, *Fractals in the Earth Science*, Barton, C.C. & P.R. LaPointe, eds., pp. 141–178, Plenum, New York.

Berkowitz, B. & Scher, H. (1997). Anomalous transport in random fracture networks, *Phys. Rev. Lett.*, Vol. 79, No. 20, 4038–4041.

Bingham, C. (1964). *Distributions on the sphere and the projective plane*, Ph.D. Dissertation, Yale University.

Bodvarsson, G.S.; Wu Y.-S.; & Zhang, K. (2003) Development of discrete flow paths in unsaturated fractures at Yucca Mountain, *J. Contam. Hydrol.*, Vol 62–63, doi:10.1016/S0169-7722(02)00177-8.

Bonnet, E.O.; Bour, O.; Odling, N.; Davy, P.; Main, I.; Cowie, P. & Berkowitz, B. (2001). Scaling of fracture systems in geologic media, *Rev. Geophys.*, Vol. 39, No. 3, 347–383.

Bour, O. & Davy, P. (1997). Connectivity of random fault networks following a power law fault length distribution, *Water Resour. Res.*, Vol. 33, 1567–1583.

Bour, O. & Davy, P. (1999). Clustering and size distributions of fault patterns: Theory and measurements, *Geophys. Res. Lett.*, Vol. 26, No. 13, 2001–2004.

Cardenas, M.B.; Slottke, D.T.; Ketcham, R.A. & Sharp, J.M. (2007). Navier-Stokes flow and transport simulations using real fractures shows heavy tailing due to eddies, *Geophys. Res. Lett.*, Vol. 34, No. L14404, doi:1029/2007/GL030554.

Chen, Z.; Neuman, S.P.; Yang, Z. & Rahman, S.S. (2000). An experimental investigation of hydraulic behavior of fractures and joints in granitic rock, *Int. J. Rock. Mech. Min. Sci.*, Vol. 37, 267–273.

Cook, A.M.; Myer, L.R.; Cook, N.G.W. & Doyle, F.M. (1990). The effect of tortuosity on flow through a natural fracture, *In: Rock Mechanics Contributions and Challenges, Proceedings of the 31st U.S. Symposium on Rock Mechanics*, W.A. Hustrulid and G.A. Johnson, eds., A.A. Balkema, Rotterdam.

Cvetkovic, V.; Painter, S.; Outters, N.; & Selroos, J.-O. (2004) Stochastic simulation of radionuclide migration in discretely fractured rock near the Äspö Hard Rock Laboratory, *Water Resour. Res.*, Vol. 40, doi:10.1029/2003WR002655.

Cvetkovic, V. (2010). Significance of fracture rim zone heterogeneity for tracer transport in crystalline rock, *Water Resour. Res.*. Vol. 46, W03504, doi:10.1029/2009WR007755.

Darcel, C.; Davy, P.; Bour, O & de Dreuzy, J.R. (2003). Connectivity properties of two-dimensional fracture networks with stochastic fractal correlation, *Water Resour. Res.*, Vol. 39, No. 10, doi:10.1029/2002WR001628.

de Dreuzy, J.R. & Erhel, J. (2003). Efficient algorithms for the determination of the connected fracture network and the solution to the steady-state flow equation in fracture networks, *Comput. Geosci.*, Vol. 29, No. 1, 107–111.

de Dreuzy, J.R.; Davy, P. & Bour, O. (2001). Hydraulic properties of two-dimensional random fracture networks following a power-law length distribution: 1. Effective connectivity, *Water Resour. Res.*, Vol. 37, No. 8, 2065–2078.

Dean, C.A. (2010). *Long-term desorption of uranium and neptunium in heterogeneous volcanic tuff materials*, Ph.D. Dissertation, University of New Mexico, Albuquerque, NM.

Fisher, R. (1953). Dispersion on a sphere, *Proc. R. Soc. Lond. Ser. A.*, 217, 295–305.

Guimerá, J. & Carrera, J. (2000). A comparison of hydraulic and transport parameters measured in low-permeability fracrtured media, *J. Contam. Hydrol.*, Vol. 41, 261-281.

Grisak, G.E. & Pickens, J.F. (1980). Solute transport through fractured media: 1. The effect of matrix diffusion, *Water Resour. Res.*, Vol. 16, No. 4, 719–730.

Gustafson, G. & Fransson, A. (2005). The use of the Pareto distribution for fracture transmissivity assessment, *Hydrogeol. J.*, Vol. 14, 15-20, doi:10.1007/s10040-005-0440-y.

Holmén, J.G. & Outters, N. (2002). *Theoretical study of rock mass investigation efficiency*, TR-02-21, Swedish Nuclear Fuel and Waste Management Company (SKB), Stockholm, Sweden.

International Atomic Energy Agency, (2011). *Nuclear Technology Review 2011*, IAEA/NTR/20011, Vienna, Austria, http://iaea.org/Publications/Reports/ntr2011.pdf.

International Atomic Energy Agency, (2001). *The Use of Scientific and Technical Results from Underground Research Laboratory Investigations for the Geological Disposal of Radioactive Waste*, IAEA-TECDOC-1243, Vienna, Austria.

Jaeger, J.C.; Cook, N.G.W. & Zimmerman, R.W. (2007). *Fundamentals of Rock Mechanics*, 4th Ed., Blackwell, Malden, MA.

Klimczak, C.; Schultz R.A.; Parashar, R. & Reeves, D.M. (2010), Cubic law with correlated aperture to length and implications for network scale fluid flow, *Hydrol. J.*, doi:10.1017/s10040-009-0572-0.

Kozubowski, T.J.; Meerschaert, M.M. & Gustafson, G. (2008), A new stochastic model for fracture transmissivity assessment, *Water Resour. Res.*, Vol. 44, No. W02435, doi:10.1029/2007WR006053.

Langmuir, D. (1997). *Aqueous Environmental Chemistry*, Prentice Hall, Upper Saddle River, New Jersey.

Long, J.C.S. & Ewing, R.C. (2004) Yucca Mountain: Earth-science issues at a geologic repository for high-level nuclear waste, *Annu. Rev. Earth Planet Sci.*, Vol. 32, 363–401, doi:10.1146/annurev.earth.32.092203.122444.

Mandelbrot, B. (1974). Intermittent turbulence in self-similar cascades: Divergence of high moments and dimension of the carrier, *J. Flud Mech.*, Vol. 62, 331–350.

Mardia, K.V. & Jupp, P.E. (2000). *Directional Statistics*, Wiley, New York.

Munier, R. (2004). *Statistical analysis of fracture data adopted for modeling discrete fracture networks – Version 2.*, Rep. R. 04-66, Swedish Nuclear Fuel and Waste Management Company (SKB), Stockholm, Sweden.

National Research Council (1978). *Geological Criteria for Repositories for High-Level Radioactive Wastes*, National Academy Press, Washington, D.C.

National Research Council (2001). *Disposition of High-Level Waste and Spent Nuclear Fuel, The Continuing Societal and Technical Challenges*, National Academy Press, Washington, D.C., ISBN 0-309-07317-0.

Neretnieks, I. (1980). Diffusion in the rock matrix: An important factor in radionuclide retention?, *J. Geophys. Res.*, Vol 85, No. B8, 4379–4397.

Neuman, S.P. (2005). Trends, prospects and challenges in quantifying flow and transport through fractured rocks, *Hydrogeol. J.*, Vol. 13, 124–147, doi:10.1007/s10040-004-0397-2.

Nuclear Energy Agency (1999). *Confidence in the Long-Term Safety of Deep Geological Repositories. Its Development and Communication*, Organisation for Economic Cooperation and Development, NEA 01809, http://www.oecd- nea.org/rwm /reports/1999/confidence.pdf.

Olson, J.E. (1993). Joint pattern development: Effects of subcritical fracture growth and mechanical crack interaction, *J. Geophys. Res.*, Vol. 98, No. B9, 12,225–12,265.

Paillet, F.L. (1998). Flow modeling and permeability estimation using borehole flow logs in heterogeneous fractured formation, *Water Resour. Res.*, Vol. 34, No. 5, 997–1010.

Painter, S.; V. Cvetkovic; & J.-O. Selroos (2002). Power-law velocity distributions in fracture networks: Numerical evidence and implications for tracer transport, *Geophys. REs. Lett.*, Vol. 29, No. 14, doi:10.1029/2002GL014960.

Pohll, G.; Hassan, A.E.; Chapman, J.B.; Papelis, C. & Andricevic, R. (1999). Modeling ground water flow and radionuclide transport in a fractured aquifer, *Ground Water*, Vol. 37, No. 5,770–784.

Pohlmann, K.; Hassan, A.E. & Chapman, J.B. (2002). Modeling density-driven flow and radionuclide transport at an underground nuclear test: Uncertainty analysis and effect of parameter correlation, *Water Resour. Res.*, Vol. 38, doi:10.1029/2001WR001047.

Pohlmann, K.; Pohll, G.; Chapman, J.; Hassan, A.E.; Carroll, R; & Shirley, C. (2004). Modeling to support groundwater contaminant boundaries for the Shoal undergound nuclear test, *Desert Research Institute Report No. 45184*.

Pohlmann, K; Ye, M.; Reves, D.M.; Zavarin, M.; Decker, D. & Chapman J. (2007) Modeling of groundwater flow and radionuclide transport at the Climax Mine sub-CAU, Nevada Test Site, *DOE/NB/26383-05, Desert Research Institute Report No. 45226*.

Qian, J.; Chen, Z.; Zhan, H. & Guan, H. (2011). Experimental study of the effect of roughness and Reynolds number on fluid flow in rough-walled single fractures: a check of the local cubic law, *Hydrol. Process.*, Vol. 25, No. 4, 614–622, doi:10.1002/hyp.7849.

Quinn, P.M.; Cheery, J.A.; & Parker, B.L. (2011). Quantification of non-Darcian flow observed during packer testing in fractured sedimentary rock, *Water Resour. Res.*, Vol 47, No. W09533, doi:10.1029/2010WR009681.

Priest, S.D. (1993). *Discontinuity Analysis of Rock Engineering*, Chapman and Hall, London.

Reeves, D.M.; Benson, D.A. & Meerschaert, M.M. (2008a). Transport of conservative solutes in simulated fracture networks: 1. Synthetic data generation, *Water Resour. Res.*, Vol. 44, No. W05401, doi:10.1029/2007WR006069.

Reeves, D.M.; Benson, D.A.; Meerschaert, M.M. & Scheffler, H.-P. (2008b). Transport of conservative solutions in simulated fracture networks: 2. Ensemble solute transport and the correspondence to operator-stable limit distributions, *Water Resour. Res.*, Vol. 44, No. W05410, doi:10.1029/2008WR006858.

Reeves, D.M.; Benson, D.A. & Meerschaert M.M. (2008c). Influence of fracture statistics on advective transport and implications for geologic repositories, *Water Resour. Res.*, Vol. 44, W08405, doi:10.1029/2007WR006179.

Reeves, D.M.; Pohlmann, K.; Pohll, G.; Ye, M. & Chapman, J. (2010). Incorporation of conceptual and parametric uncertainty into radionuclide flux estimates from a fractured granite rock mass, *Stoch. Environ. Res. Risk Assess.*, doi:10.1007/s00477-010-0385-0.

Renshaw, C.E. (1995). On the relationship between mechanical and hydraulic apertures in rough walled fractures, *J. Geophys. Res.*, Vol. 100, No. B12, 24,629–24,363.

Renshaw, C.E. (1999). Connectivity of joint networks with power law length distributions, *Water Resour. Res.*, Vol. 35, No. 9, 2661–2670.

Rives, T.M.; Razack, M.; Petit, J.-P. & Rawnsley, K.D. (1992). Joint spacing: Analogue and numerical simulations, *J. Struct. Geol.*, Vol. 14, 925–937.

Robinson, B.A.; Li, C. & Ho, C.K. (2003). Performance assessment model development and analysis of radionuclide transport in the unsaturated zone, Yucca Mountain, Nevada, *J. Contam. Hydrol.*, Vol. 62-63, doi:10.1016/S0169-7722(02)00166-3.

Ross, S.M. (1985). *Introduction to Probility Models*, 3rd Ed., Academic Press, Orlando, FL.

Schertzer, D. & Lovejoy, S. (1987). Physical modeling and analysis of rain and clouds by anisotropic scaling multiplicative processes, *J. Geophys. Res.*, Vol. 85, No. D8, 9693–9714.

Schwartz, F.W.; Smith, L.; & Crowe, A.S. (1983). A stochastic analysis of macroscopic dispersion in fractured media, *Water Resour. Res.*, Vol. 19, No.5, 1253–1265.

Segall, P. & Pollard, D.D. (1983). Joint formation in granitic rock in the Sierra Nevada, *Geol. Soc. Am. Bull.*, Vol. 94, 563–575.

Smith, P.A.; Alexander, W.R.; Kickmaier, W.; Ota, K.; Frieg, K. & McKinley, I.G. (2001). Development and testing of radionuclide transport models for fractured rock: examples for the Nagra/JNC Radionuclide Migration Programme in the Grimsel Test Site, Switzerland, *J. Contam. Hydrol.*, Vol. 47, No. 2-4, 335–348.

Smith, L. & Schwartz, F.W. (1984). An analysis of the influence of fracture geometry on mass transport in fractured media, *Water Resour. Res.*, Vol. 20, No. 9, 1241–1252.

Snow, D.T. (1965). *A parallel plate model of fractured permeability media*, Ph.D. Dissertation, University of California, Berkeley.

Stigsson, M.; Outters, N. & Hermanson, J. (2001). *Äspö Hard Rock Laboratory, Prototype Repository Hydraulic DFN Model no. 2*, IPR-01-39, Swedish Nuclear Fuel and Waste Management Company (SKB), Stockholm, Sweden.

Sudicky, E.A. & Frind, O.E. (1982). Contaminant transport in fractured porous media: Analytical solutions for a system of parallel fractures, *Water Resour. Res.*, Vol. 18, No. 6, 1634–1642.

Terzaghi, R. (1965). Sources of error in joint surveys, *Geotechnqiue*, Vol. 15, No. 3, 287–304.

Twiss, R.J. & Moores, E.M. (2007). *Structural Geology*, 2nd Ed., W.H. Freeman, New York.

Winberg, A; Andersson P.; Hermanson J.; Byegard J.; Cvetkovic, V.; & Birgersson, L. (2000). *Aspo Hard Rock Laboratory, Final Report of the First Stage of the Tracer Retention Understanding Experiments*, Technical Report TR-00-07, Swedish Nuclear Fuel and Waste Management Co. (SKB), Stockholm, Sweden.

Wood, A.T.A. (1994). Simulation of the Von Mises distribution, *Commun. Stat.-Sim.*, Vol. 21, No. 1, 157–164.

Zhang Y.; Baeumer, B. & Reeves, D.M. (2010). A tempered multiscaling stable model to simulate transport regional-scale fractured media, *Geophys. Res. Lett.*, Vol. 37, No. L11405, doi:10.1029/2010GL043609.

Zheng, C. & Bennett, G.D. (2002). *Applied Contaminant Transport Modeling*, John Wiley and Sons, Inc., New York.

Zimmermann, R.W. & Bodvarsson, G.S. (1996). Hydraulic conductivity of rock fractures, *Trans. Porous Media*, Vol. 23, 1–30.

Diffusion of Radionuclides in Concrete and Soil

Shas V. Mattigod[1], Dawn M. Wellman[1], Chase C. Bovaird[1],
Kent E. Parker[1], Kurtis P. Recknagle[1], Libby Clayton[2] and Marc I. Wood[3]

[1]*Pacific Northwest National Laboratory, Richland, WA*
[2]*BP Exploration, Anchorage, AK*
[3]*CH2M Hill Plateau Remediation Company, Richland, WA*
USA

1. Introduction

One of the methods being considered for safely disposing of low-level radioactive wastes (LLW) is to encase the waste in concrete. Such concrete encasement would contain and isolate the waste packages from the hydrologic environment and would act as an intrusion barrier. The current plan for waste isolation consists of stacking low-level waste packages on a trench floor, surrounding the stacks with reinforced steel, and encasing these packages in concrete. These concrete-encased waste stacks are expected to vary in size with maximum dimensions of 6.4 m long, 2.7 m wide, and 4 m high. The waste stacks are expected to have a surrounding minimum thickness of 15 cm of concrete encasement. These concrete-encased waste packages are expected to withstand environmental exposure (solar radiation, temperature variations, and precipitation) until an interim soil cover or permanent closure cover is installed, and to remain largely intact thereafter.

Any failure of concrete encasement may result in water intrusion and consequent mobilization of radionuclides from the waste packages. The mobilized radionuclides may escape from the encased concrete by mass flow and/or diffusion and move into the surrounding subsurface environment. Therefore, it is necessary to assess the performance of the concrete encasement structure and the ability of the surrounding soil to retard radionuclide migration. The retardation factors for radionuclides contained in the waste packages can be determined from measurements of diffusion coefficients for these contaminants through concrete and fill material. Some of the mobilization scenarios include 1) potential leaching of waste form before permanent closure cover is installed; 2) after the cover installation, long-term diffusion of radionuclides from concrete waste form into surrounding fill material; 3) diffusion of radionuclides from contaminated soils into adjoining concrete encasement and clean fill material. Additionally, the rate of diffusion of radionuclides may be affected by the formation of structural cracks in concrete, the carbonation of the buried waste form, and any potential effect of metallic iron (in the form of rebar) on the mobility of radionuclides.

The radionuclides iodine-129 ([129]I), technetium-99 (^{99}Tc), and uranium-238 ([238]U) are identified as long-term dose contributors in LLW (Mann et al. 2001; Wood et al. 1995). Because of their anionic nature in aqueous solutions, [129]I, ^{99}Tc, and carbonate-complexed

^{238}U may readily leach into the subsurface environment (Serne et al. 1989, 1992a, b, 1993, and 1995). The leachability and/or diffusion of radionuclide species must be measured to assess the long-term performance of waste grouts when contacted with vadose-zone pore water or groundwater.

Although significant research has been conducted on the design and performance of cementitious waste forms, the current protocol conducted to assess radionuclide stability within these waste forms has been limited to the Toxicity Characteristic Leaching Procedure, Method 1311 Federal Registry (Environmental Protection Agency [EPA] 1992) and ANSI/ANS-16.1 leach test (American National Standards Institute [ANSI] 1986). These tests evaluate the performance under water-saturated conditions and do not evaluate the performance of cementitious waste forms within the context of waste repositories which are located within water-deficient vadose zones. Moreover, these tests assess only the diffusion of radionuclides from concrete waste forms and neglect evaluating the mechanisms of retention, stability of the waste form, and formation of secondary phases during weathering, which may serve as long-term secondary hosts for immobilization of radionuclides.

The results of recent investigations conducted under arid and semi-arid conditions (Al-Khayat et al. 2002; Garrabrants et al. 2002, 2003, and 2004; Gervais et al. 2004; Sanchez et al. 2002, 2003) provide valuable information suggesting structural and chemical changes to concrete waste forms which may affect contaminant containment and waste form performance. However, continued research is necessitated by the need to understand the mechanism of contaminant release; the significance of contaminant release pathways; how waste form performance is affected by the full range of environmental conditions within the disposal facility; the process of waste form aging under conditions that are representative of processes occurring in response to changing environmental conditions within the disposal facility; the effect of waste form aging on chemical, physical, and radiological properties, and the associated impact on contaminant release. Recent reviews conducted by the National Academies of Science recognized the efficacy of cementitious materials for waste isolation, but further noted the significant shortcomings in our current understanding and testing protocol for evaluating the performance of various formulations. The objective of our study was to measure the diffusivity of Re, Tc and I in concrete containment and the surrounding vadose zone soil. Effects of carbonation, presence of metallic iron, and fracturing of concrete and the varying moisture contents in soil on the diffusivities of Tc and I were evaluated.

2. Concrete composition and fabrication of test specimens

The concrete composition for the burial encasement was indicated in Specification for Concrete Encasement for Contact-Handled LLW. This specification was used as the basis to prepare concrete for fabrication of test specimens. Details of concrete composition and test specimen fabrication has been described in a technical report (Wellman et al. 2006).

2.1 Specified concrete composition for encasement

The specified composition includes sulfate-resistant Portland Type I or Type II cement, a pozzolanic material (Class F fly ash), fine and coarse aggregates, and steel fiber. Additional specifications include a water-to-cement ratio of 0.4 and an air content 6.0 ± 1.5%. The nominal proportions and material specifications based on this initial design are listed in the material specifications for field mix column in Table 1.

A laboratory-scale concrete mixture (Table 1) was prepared based on specifications for Portland Type I and Type II cement. Because of the required small dimensions of laboratory test specimens, the size of the coarse aggregate and the dimensions of the steel fiber were proportionately reduced. This was accomplished by decreasing the 2-cm (~0.75 in.) coarse aggregate size in the original specification to a particle size ranging from 2.83 mm to 2 mm in the laboratory mix. The aggregate mixture passing a 7-mesh sieve and retained on a 10-mesh sieve met this particle size specification. The scaled-down steel fibers used in the laboratory mix consisted of Bekaert Dramix brand deformed steel wire fibers that were cut to a nominal length of 8 mm (0.31 in). The deformed end portions were retained for use in the concrete mixture and the straight middle section of the fiber was discarded. Based on these modifications, a concrete mix was prepared that consisted of Portland Cement (Type I & II, ASTM C-150 compliant), Class F fly ash, scaled-down coarse aggregate, fine aggregate, scaled-down deformed steel fiber, and a water-entraining agent (PolyHeed® 997). The water-entraining agent was included in the mix to facilitate the workability of the concrete. The volumes of the PolyHeed® 997 and the air-entraining agent, MB-AE™ 90, were not included in the normalization calculations because of their negligible contribution to the overall mix volume. The material specification and composition for the laboratory-scale concrete mixture is given in Table 1.

Material	Material Specifications for Field Mix	Normalized Laboratory Design	Material Specifications Used in Revised Laboratory Mix Comparison
Cement	Portland Type I or Type I/II sulfate-resistant cement	0.27	Portland Type I & II
Fly Ash	Class F fly ash; nominal 15% of cement by volume	0.04	Class F fly ash; nominal 20% of cement by volume
Coarse Aggregate	No. 676 or equivalent (3/4" nominal size)	0.04	Sieve size +7 to -10 (2.83 – 2 mm size)
Fine Aggregate	Sand	0.51	Sand -10 sieve size (< 2 mm)
Water	Nominal water-to-cement ratio: 0.4	0.10	Water-to-cement ratio: 0.5
Steel Fiber	Deformed Type I, nominal length 2.5 – 3.8 cm (1 – 1.5")	0.04	Deformed, nominal length 8 mm (0.32 in)
PolyHeed® 997	N/A	0.00375	Water-entraining agent
Air Content	6.0±1.5%	6.0±1.5%	N/A

Table 1. Laboratory-Scale Material Specification and Composition

2.1.1 Concrete mold design

The concrete molds for casting specimens were fabricated from Schedule 40 polyvinyl chloride (PVC) piping material. Gaskets were glued to the bottom of the molds and leak tested before use. The PVC forms were pre-treated with form release, a liquid that allows the concrete specimen to release easily from the mold. The first treatment was applied 3 days prior, and the second treatment was applied a few hours before wet concrete was added to the molds.

2.1.2 Concrete mix and specimen preparation

Concrete monoliths were prepared with mix components added in this order: water, steel (if applicable), coarse aggregate, fine aggregate, fly ash, cement, PolyHeed® 997, and MB-AE™ 90. The concrete was mixed on medium speed using a Hobart 3-speed bench top mixer in a 4 L steel bowl. The PVC molds were filled in the vertical positions. After filling, the molds were lightly tapped on the laboratory bench until a significant decrease in the release of air bubbles was observed. The forms were stored in plastic bags with damp paper towels to provide moisture while the concrete set.

2.2 Concrete-soil diffusion tests

The objective of half-cell tests was to examine the rate of diffusion of key, long-lived, mobile contaminants in unsaturated Hanford Site vadose zone sediment in contact with concrete monolith. These experiments were conducted using half cells of Trench 8 soil (~4-cm diameter and ~10-cm long) in contact with concrete monolith (~4-cm diameter and ~4-cm long). The soil was oven dried, then brought up to 4, 7, or 15% soil moisture content with 18 MΩ cm distilled de-ionized water. Moisture content analysis was performed and the nominal moisture content for 4% cores was 1%, 7% cores were found to be 4%, and 15% were found to be 11%.

For spiked concrete tests, iodine, technetium or rhenium (as Tc surrogate) were first added to water and then mixed into the dry ingredients. For soil spiked tests, the contaminants were first added to water and then added oven-dried soil. Sampling of soil half-cells was conducted at appropriate time intervals by extruding and taking transverse sections of the sediment at about 0.5-cm intervals for the first 4 to 6 cm from the interface and at 1-cm intervals for the remaining part of the half cell. The soil samples were weighed and extracted with either deionized distilled water, and the extracts were analyzed for stable iodide, rhenium and technetium-99 using ICP-MS. The diffusion profile in concrete half-cells were determined by dry-slicing the cells with a circular diamond saw. The resulting thin slices were then crushed and extracted with deionized distilled water and the solutions were analyzed for concentrations of the contaminant of interest.

3. Diffusion coefficient calculations

The radionuclide diffusivity is defined by Fick's equation as

$$J = - D_w \, dC_w / dx \tag{1}$$

where, J = flux of radionuclide at a given point, D_w = the diffusivity of water-based radionuclide concentration, and C_w = the radionuclide concentration in the pore water.

Ignoring the possibility that radionuclides might be present in the air within the unsaturated sediment, and recognizing that in the case of a two-phase system (water and soil), Equation 1 can be used in performing a mass balance over a small volume leading to the equation (Crank 1975)

$$dC_w/dt = D_w/\theta * (d^2C_w/dx^2) \tag{2}$$

where θ = the volume fraction water in the soil's pore space. However, the slope on the probit plot provides the diffusivity that solves the equation for diffusion in a homogeneous single phase medium,

$$dC/dt = D*(d^2C/dx^2) \tag{3}$$

The solution for this diffusion equation is provided by

$$C/C_0 = \tfrac{1}{2} \, erfc \, (x/(\sqrt{4Dt})) \tag{4}$$

Where, C/C_0 is the normalized concentration, and erfc is the error function (Crank 1975). The diffusivities in the soil were calculated using a probit analysis approach. The details of the probit analysis are provided in Finney (1971). This technique allows the transformation of a sigmoid curve of concentrations, normalized with respect to the initial concentration (C_0) as a function of diffusion distance (x) produced in a half-cell diffusion experiment to a linear plot. The transformed equation is,

$$Probit \, (p) = \Phi^{-1} \, (p) = a - x/\sqrt{(2Dt)} \tag{5}$$

When probit transform of C/C_0 is plotted against x, the resulting slope b = $1/\sqrt{(2Dt)}$ can used to calculate the diffusivity (D) as D= $1/(2b^2t)$, where t is the sampling time. This approach has been used previously to determine diffusivity in half-cell diffusion experiments (Brown et al. 1964. Lamar 1989).The diffusion coefficient D_w that was calculated from the D obtained from the probit plot based on concentrations in the pore water must then be multiplied by θ. Detailed statistical analysis (Abdel Rahman and Zaki 2011) was not conducted to assess the uncertainities associated with slopes from the linear regression and the calculated diffusion coefficient values.

4. Effects of carbonation and metallic iron on the diffusion of technetium and iodine in soils and concrete

Diffusion experiments were conducted using carbonated (by immersion in sodium carbonate solution) concrete-soil half cells. Soil half-cell specimens were spiked with I and Tc to achieve a measurable diffusion profile in the concrete part of the half-cell. The goal of this test was to examine the effect of carbonation and metallic iron addition on diffusion of I and Tc into concrete.

Typical diffusion profiles for Tc and I in concrete and the corresponding Probit plots are shown in figures 1 and 2.

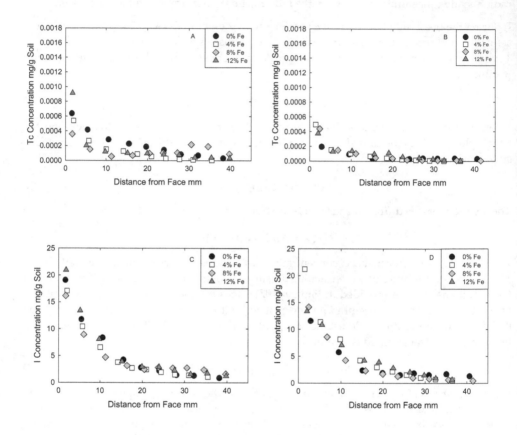

Fig. 1. Concrete half-cell concentration profiles as a function of iron content:
Top left: Tc concentration for non-carbonated concrete at 4% soil moisture;
Top right: Tc concentration for carbonated concrete at 4% soil moisture;
Bottom left: I concentration for non-carbonated concrete at 4% soil moisture;
Bottom right: I concentration for carbonated concrete at 4% soil moisture.

The results indicated diffusivities ranging from 1×10^{-10} to 1.9×10^{-9} cm²/s for Tc, and 1.3×10^{-10} to 2.3×10^{-9} cm²/s for I (Table 2). The highest Tc and I diffusivities were observed in all uncarbonated, Fe-free concrete cores contacting spiked soils at all three moisture contents (4%, 7%, and 15%). However, the diffusivities of both Tc and I (except in one case) were significantly attenuated in all carbonated concrete cores. The reduction of Tc diffusivities ranged from 55% to 72%. Meanwhile, I diffusivities were reduced by 61% at soil moisture content of 4% and by 58% at soil moisture content of 15%. However, I diffusivity showed an anomalous increase of ~38% in an uncarbonated, Fe-free concrete core in contact with a soil core with 7% moisture content.

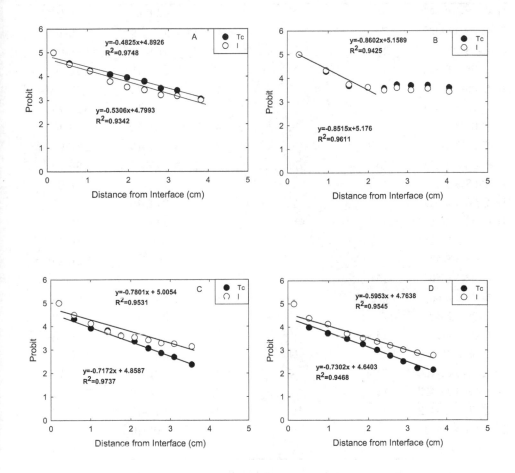

Fig. 2. Probit plots for concrete cores. Top left: 4% moisture content, no Fe, no carbonation; Top right: 4% moisture content, no Fe, carbonated; Bottom left: 4% moisture content, 4% Fe, no carbonation; Bottom right: 4% moisture content, 4% Fe, carbonated.

Similar attenuation in Tc and I diffusivities were also observed in carbonated concrete half cells containing various quantities of Fe. The diffusivities of Tc showed reduction typically ranging from 25 to 81%, except enhanced diffusivities (~40 - 30%) were observed in two concrete cores containing 4% Fe and in contact with soil with 4 and 15% moisture contents. In these same cores, similar increases in I diffusivities were also observed (10% - 6%). However all other Fe-containing carbonated concrete cores exhibited reduction in I diffusivities that ranged from 9% to 76%.

Core ID	Initial Moisture Content (Wt %)	Carbonation	Fe (wt %)	Tc Diffusivity (cm²/s)	I Diffusivity (cm²/s)
Tc-C-08-3-0-325	4	N	0	1.7E-09	1.4E-09
Tc-C-08-3-0-329	7	N	0	1.9E-09	4.0E-10
Tc-C-08-3-0-330	15	N	0	1.3E-09	1.2E-09
Tc-C-08-3-0-332	4	Y	0	5.3E-10	5.4E-10
Tc-C-08-3-0-333	7	Y	0	5.4E-10	5.5E-10
Tc-C-08-3-0-334	15	Y	0	5.9E-10	5.1E-10
Tc-C-08-3-4-350	4	N	4	7.7E-10	6.5E-10
Tc-C-08-3-4-351	7	N	4	6.9E-10	2.3E-09
Tc-C-08-3-4-353	15	N	4	3.6E-10	7.9E-10
Tc-C-08-3-4-357	4	Y	4	7.4E-10	1.1E-09
Tc-C-08-3-4-359	7	Y	4	2.6E-10	4.2E-10
Tc-C-08-3-4-360	15	Y	4	7.8E-10	5.4E-10
Tc-C-08-3-8-401	4	N	8	1.9E-10	7.1E-10
Tc-C-08-3-8-402	7	N	8	(a)	5.6E-10
Tc-C-08-3-8-403	15	N	8	(a)	6.0E-10
Tc-C-08-3-8-407	4	Y	8	5.1E-10	4.9E-10
Tc-C-08-3-8-409	7	Y	8	2.2E-10	3.4E-10
Tc-C-08-3-8-410	15	Y	8	3.6E-10	3.4E-10
Tc-C-08-3-12-425	4	N	12	(a)	4.3E-10
Tc-C-08-3-12-426	7	N	12	2.3E-10	4.0E-10
Tc-C-08-3-12-427	15	N	12	4.7E-10	7.7E-10
Tc-C-08-3-12-432	4	Y	12	1.1E-10	1.3E-10
Tc-C-08-3-12-433	7	Y	12	1.0E-10	3.3E-10
Tc-C-08-3-12-435	15	Y	12	4.4E-10	3.1E-10

(a) Probit analysis could not be performed due to poorly defined diffusion profile

Table 2. Diffusivity of Tc and I in concrete half cells

5. Diffusion of iodine, rhenium, and technetium from soil into fractured concrete

A set of diffusion experiments were conducted to examine the effect of fracturing of spiked encasement concrete on the diffusion of radionuclides into soil. Spiked and cured concrete monoliths were encased in shrink wrap and struck with a hammer to prevent the formation of rubble. Each fractured core possessed a single fracture extending the length of the core, perpendicular to the concrete-sediment interface.

In one set of experiments, soil half-cell specimens were spiked with I and Re to achieve a measurable diffusion in the fractured concrete part of the half-cell. The diffusion tests were conducted under unsaturated conditions at 4%, 7%, and 15% (moisture content by weight) with or without Fe addition (4 wt %) and carbonation. In a second set of experiments, Tc-spiked soil half cells at a fixed moisture content were contacted with fractured concrete half cells.

Typical probit plots are shown in Figure 3 and the data is listed in Table 3. Overall, the calculated diffusivities for Re ranged from 3.8×10^{-12} to 2×10^{-9} cm^2/s and for I ranged from 1.3×10^{-10} to 2.3×10^{-9} cm^2/s in fractured concrete. The highest Re and I diffusivities were observed in fractured concrete cores that were in contact with spiked soils with 15% moisture content. Effects of carbonation in enhancing diffusion are noticeable at higher moisture contents (7 and 15%) of the contacting soil. Additions of 4% by weight Fe did not noticeably affect the diffusivities of Re and I in fractured concrete.

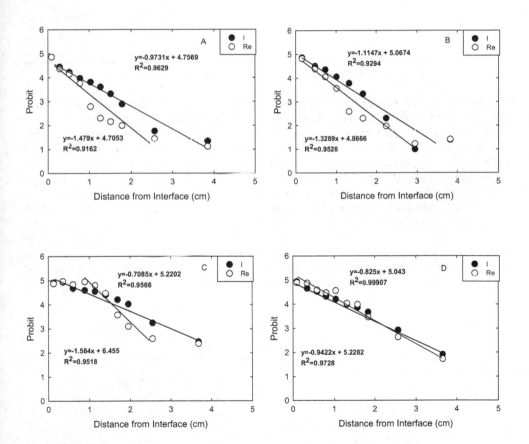

Fig. 3. Probit plots for concrete cores. Top left: 4% moisture content, no Fe, no carbonation; Top right: 4% moisture content, 4% Fe, no carbonation; Bottom left: 4% moisture content, no Fe, carbonated; Bottom right: 4% moisture content, 4% Fe, carbonated half cells.

Core ID	Moisture Content wt%	Carbonation	Fe (wt %)	Re Diffusivity (cm²/s)	I Diffusivity (cm²/s)
C-5-0-2	4	N	0	1.36E-10	3.13E-10
C-5-4-26	4	N	4	1.68E-10	2.39E-10
C-5-0-1	4	Y	0	1.18E-10	5.90E-10
C-5-4-21	4	Y	4	3.34E-10	4.35E-10
C-5-0-7	7	N	0	3.75E-12	1.25E-10
C-5-4-27	7	N	4	8.88E-11	1.93E-10
C-5-0-5	7	Y	0	2.60E-10	2.77E-10
C-5-4-23	7	Y	4	1.69E-10	4.00E-10
C-5-0-10	15	N	0	2.03E-09	2.33E-09
C-5-4-30	15	N	4	4.02E-10	1.76E-09
C-5-0-6	15	Y	0	6.90E-10	5.10E-10
C-5-4-24	15	Y	4	6.72E-10	1.21E-09

Table 3. Diffusivity data for Re-I in fractured concrete half cells

In a second set of experiments, Tc-spiked soil half cells at a fixed moisture content were contacted with fractured concrete half cells. The resulting probit analysis for Tc diffusivity in concrete cores is listed in Table 4. These values ranged from 3.1×10^{-11} cm²/s to 3.6×10^{-10} cm²/s. The data indicated that carbonation in all cases enhanced Tc diffusivity in concrete cores. Significant increases in Tc diffusivities were noted when cores with higher concentrations of Fe (8% and 12%) were carbonated. These data indicate that carbonation of Fe containing concrete cores may enhance micro-cracking of concrete resulting in an increase in Tc diffusivity. Similar phenomena have been noted in previous studies in which Fe containing carbonated concrete cores was in contact with Tc-spiked soil cores with 4% moisture content. Also, the beneficial effect of Fe on reducing Tc diffusivity in non-carbonated specimens is not observable until the Fe content is at least 8% by mass. These data indicated that diffusivity in concrete is controlled by carbonation and Fe content and the presence of a single fracture had no measureable effect.

Core ID	Carbonation	Fe (wt %)	Tc Diffusivity (cm²/s)
Tc-C-10-5-0-101	N	0	1.1E-10
Tc-C-10-5-0-102	Y	0	1.4E-10⁰
Tc-C-10-5-4-105	N	4	2.1E-10
Tc-C-10-5-4-106	Y	4	3.6E-10
Tc-C-10-5-8-107	N	8	3.1E-11
Tc-C-10-5-8-108	Y	8	1.9E-10
Tc-C-10-5-12-110	N	12	8.4E-11
Tc-C-10-5-12-111	Y	12	2.1E-10

Table 4. Diffusivity of Tc in fractured concrete cores

6. Diffusion of iodine and rhenium from fractured concrete into soil

Two sets of diffusion experiments were initiated using fractured concrete-soil half cells. The objective was to examine the diffusion from spiked and fractured concrete cores into soil half cells. In one set, Re and I spiked concrete specimens were prepared with and without carbonation and additions of 0, 4, 8, and 12% Fe. After curing, these concrete specimens were fractured and half cells were prepared by placing these specimens in contact with soil at 4, 7, and 15% initial moisture content. Typical probit plots are shown in Figure 4 and the results of probit analysis for these sets of diffusion data are shown in Table 5.

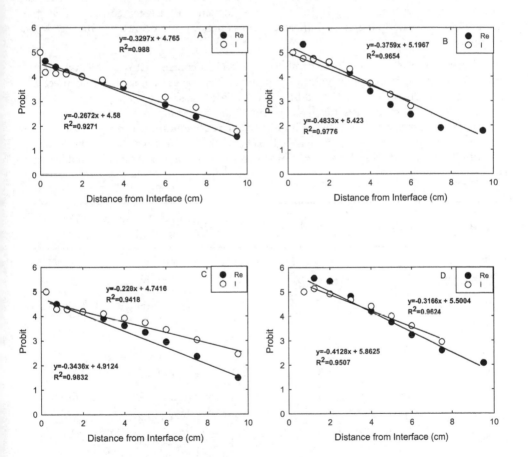

Fig. 4. Probit analysis for Re and I in soil cores. Top left: 4% moisture content, no Fe, no carbonation; Top right: 4% moisture content, no Fe, carbonated; Bottom left: 4% moisture content, 4% Fe, no carbonation; Bottom right: 4% moisture content, 4% Fe, carbonated.

The calculations showed that at 4 % moisture content, with no iron present, carbonation reduced the diffusivities of both Re and I in soils, roughly by half. The addition of 4% by mass Fe into un-carbonated concrete did not significantly affect Re and I diffusivities in soil. In carbonated specimens, adding 4% Fe seemed to increase only I diffusivity without significantly affecting Re diffusivity. When the concentrations of Fe in concrete cores were increased to 8%, significantly enhanced diffusivities in soils were observed. For instance, in soils in contact with un-carbonated concrete cores, the presence of 8% Fe in the concrete increased Re and I diffusivities by ~17 and 36%, respectively. With carbonation, however, the diffusivities of both Re and I more than doubled as compared to carbonated specimens with no Fe. Similarly, increasing the Fe content to 12% by mass in both un-carbonated and carbonated concrete increased Re diffusivities soils by about a third as compared to soils contacting cores with no added Fe. In soils contacting similarly treated concrete cores, the I diffusivities did not change significantly (carbonation) or decreased by ~20% (no carbonation).

In soil cores with higher moisture content (7%) and in contact with carbonated concrete cores containing no iron, the diffusivity of Re was ~20% lower than in soils in contact with un-carbonated concrete cores. The addition of 4% Fe by mass to carbonated concrete cores increased Re and I soil diffusivities by ~1.5 and ~3 times as compared to diffusivities in soil cores contacting concrete half cells with no Fe content. Rhenium diffusivity in soil cores in contact with noncarbonated concrete half cells with 12% Fe was about an order of magnitude lower than in soil cores in contact with similar concrete half cells containing no Fe.

Core ID	MC (Wt %)	Carbonation	Fe (wt %)	Re Diffusivity (cm^2/s)	I Diffusivity (cm^2/s)
C-08-5-0-501	4	N	0	2.2E-09	3.3E-09
C-08-5-0-502	7	N	0	2.1E-07	NA[a]
C-08-5-0-504	4	Y	0	1.0E-09	1.7E-09
C-08-5-0-505	7	Y	0	1.7E-07	1.2E-07
C-08-5-4-526	4	N	4	2.0E-09	4.5E-09
C-08-5-4-530	4	Y	4	1.4E-09	2.3E-09
C-08-5-4-531	7	Y	4	2.6E-07	3.6E-07
C-08-5-8-552	4	N	8	2.5E-08	4.4E-08
C-08-5-8-555	4	Y	8	2.4E-09	3.8E-09
C-08-5-12-576	4	N	12	1.5E-09	2.6E-08
C-08-5-12-580	4	Y	12	1.4E-09	1.7E-09

[a] Probit analysis could not be performed due to poorly defined diffusion profile

Table 5. Diffusivity analysis Re and I in soil half cells

In soil cores (7% moisture content) contacting Fe-free concrete with and without carbonation, the Re and I diffusivities were more than an order of magnitude higher than in soils containing 4% moisture content.

At the end of the experiment lasting over a year, Re and I diffusion in soil cores with 15% moisture content had proceeded to the degree that no distinct concentration gradients were present; therefore, diffusivity values could not be ascertained. These data suggest that diffusivities in soil are affected mainly by its moisture content. State of carbonation, Fe content and the presence of fracture in source- concrete do not significantly affect soil diffusivities except mass transfer of contaminants.

7. Modeling the diffusion of iodine and technetium from a reservoir through a concrete enclosure and into surrounding soil

To model the diffusion of iodine and technetium through 6 inches of concrete into the surrounding soil, pairs of high- and low-diffusivity values were selected from the experimental data. The values selected are listed in Table 6.

Diffusing Component	Material	Low Diffusivity, cm²/s	High Diffusivity, cm²/s
Iodine	Concrete	1.30E-10	2.30E-09
	Soil	1.70E-09	2.50E-08
Tc	Concrete	3.10E-11	3.60E-10
	Soil	1.30E-08	5.80E-08

Table 6. Diffusivity values for concrete and soil used in the simulations

7.1 Modeling approach

The computational fluid dynamics (CFD) code, STAR-CD[1] was used for the calculations. STAR-CD is a commercial CFD code that solves the finite volume formulations for conservation of mass, momentum, and energy for general-purpose thermal-fluids simulations. STAR-CD was used to simulate the iodine and technetium species diffusion through the concrete encasement box and into the soil using the analogy of thermal conduction,

$$\frac{dT}{dt} = \frac{k}{\rho C_p} \frac{d^2T}{dx^2} \tag{6}$$

and mass diffusion:

$$\frac{dC}{dt} = D \frac{d^2C}{dx^2} \tag{7}$$

---

[1] STAR-CD, *Version 4.14 Methodology* 2010: Computational Dynamics Ltd.

In the heat conduction equation (6) T is the absolute temperature in Kelvin, ρ is the density, and C_p is the specific heat of the material. In the mass diffusion equation (7) C is the specie concentration and D is the specie diffusivity. In STAR-CD we can simulate the specie diffusion through the porous concrete and soil by setting the coefficients of (6) and (7) as equal and solving the thermal conduction problem through the two materials.

The following assumptions were made for performing the simulations:

- Constant specie concentration reservoir exists adjacent to inside wall of concrete encasement box.
- 1-dimensional diffusion through encasement box wall and into the surrounding soil.
- No additional resistance between concrete and soil to decrease the specie diffusion.

A one-dimensional (1-D) finite volume model was constructed to represent the species transport through the concrete encasement wall and into the soil. A 1-D rectangular mesh was used for transport through the encasement wall. As the species exit the concrete and enter the surrounding soil, the environment is more cylindrical in nature as the species are free to diffuse radially into the soil. Thus the mesh for the concrete wall was mated to a cylindrical mesh extending out into the surrounding soil as shown schematically in Figure 5. The model domain is shown outlined with a dashed line in the figure. In the model, the concrete wall is assumed to be in contact with a constant specie reservoir (at location 1 in the figure). The contact between concrete and soil (at location 2 in the figure) is assumed to be perfect with no additional resistance there for specie diffusion. The outer radius of the model was established far away from the reservoir such that specie concentration remained zero for the times examined (1000, 3000, 10000, and 30000 years). Simulations were performed for diffusion of Iodine and Tc under low-diffusivity and high-diffusivity conditions (Table 6).

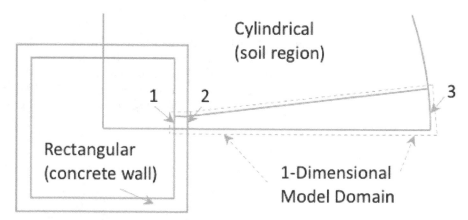

Fig. 5. Plan-view schematic of the model domain (dashed outline) where the rectangular concrete encasement wall is in contact with the surrounding soil region into which the species diffuse radially. 1. Concrete wall in contact with specie reservoir. 2. Concrete in contact with soil. 3. Outer radius is established far enough away from the reservoir that species concentration is zero at long times (i.e., 30,000 years).

7.2 Results

The results of the simulations are listed in 7 and 8 and the diffusion profiles are shown in Figures 6 through 9.

The normalized concentrations (C/C_0) at the concrete (~15 cm thick) soil interface as a function of time are listed in Table 7. The data indicates that under low diffusivity conditions, the C/C_0 for both I and Tc at all simulated times are very low (0 - 0.07). Under high diffusivity conditions, the C/C_0 values for Tc ranged from 0 - 0.03, whereas, iodine had C/C_0 values ranging from 0.05 - 0.22 indicating deeper penetration into concrete.

Diffusivity	Time (Y)			
	1,000	3,000	10,000	30,000
	C/C_0			
I Low	0.00	0.00	0.02	0.07
I high	0.05	0.10	0.17	0.22
Tc Low	0.00	0.00	0.00	0.00
Tc High	0.00	0.01	0.02	0.03

Table 7. Normalized concentration values for I and Tc at the concrete-soil interface

The depths of penetration, defined as C/C_0 of ~0.005 for various simulated time periods, are listed in Table 8. Under low diffusivity conditions, iodine at the end of 30,000 years is predicted to penetrate about 69 cm into soil (Table 8 and Figure 6), whereas Tc appears to be confined within the encasing of ~15-cm thick concrete (Table 8 and Figure 8). Under higher diffusivity conditions, the simulations indicated higher depths of penetration of I and Tc into soil. For instance, iodine is predicted to penetrate from ~ 39 - 398 cm into the soil for the time period ranging from 1,000 - 30,000 years (Table 8, Figure 7). Technetium under high diffusivity conditions is predicted to penetrate ~112 -286 cm into the soil at the end of 10,000 and 30,000 years respectively (Table 8, Figure 9).

Diffusivity	Time (Y)			
	1,000	3,000	10,000	30,000
	Depth of Soil Penetration (cm) at C/C_0 = ~0.005			
I Low	0	0	18	69
I high	39	99	215	398
Tc Low	0	0	0	0
Tc High	0	0	112	286

Table 8. Depth of penetration of I and Tc into soil as a function of time

The experimental data indicates that low diffusivity for I is engendered when carbonated encasing concrete is surrounded by soil with very low moisture contents. For Tc, low diffusivity is observed when carbonated and Fe-containing concrete is in contact with very low moisture content sediments. Comparatively, higher diffusivities of I and Tc are typically observed when higher moisture content soil is in contact with uncarbonated, Fe-free encasing concrete.

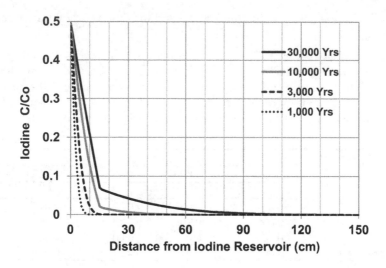

Fig. 6. Iodine concentration profiles for the low diffusivity case. Zero to 15 cm represents concentrations within the concrete encasement wall. Fifteen cm beyond are concentrations in the surrounding soil. $C/C_0 = 0.005$ was about 69 cm into the soil at 30,000 years.

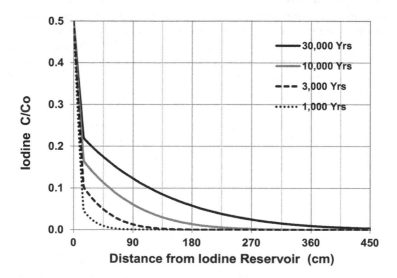

Fig. 7. Iodine concentration profiles for the high diffusivity case. Zero to 15 cm represents concentrations within the concrete encasement wall. Fifteen cm and beyond are concentrations in the surrounding soil. $C/C_0 = 0.005$ was about 398 cm into the soil depth at 30,000 years.

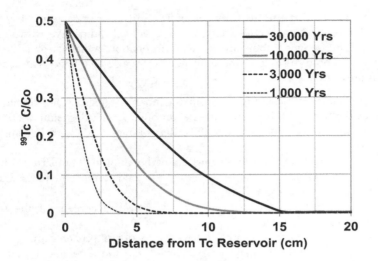

Fig. 8. Technetium concentration profiles for the low diffusivity case. Zero to 15 cm represents concentrations within the concrete encasement wall. Fifteen cm and beyond are concentrations in the surrounding soil. $C/C_0 = 0.005$ was about to the outer edge of the concrete encasement (15 cm total penetration depth) at 30,000 years.

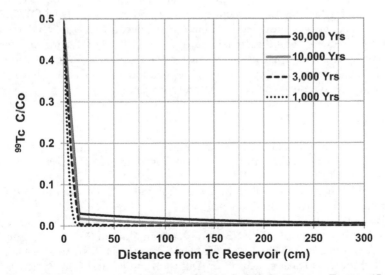

Fig. 9. Technetium concentration profiles for the high diffusivity case. Zero to 15 cm represents concentrations within the concrete encasement wall. Fifteen cm and beyond are concentrations in the surrounding soil. $C/C_0 = 0.005$ was about 286 cm into the soil at 30,000 years.

8. Conclusions

Tests were conducted to measure the diffusion of Tc, Re (as a surrogate for Tc), and I in unsaturated sediments and in concrete that encases low-level radioactive wastes. In these tests, effects of varying sediment moisture content, carbonation, metallic iron addition, and fracturing of concrete on diffusivity of these nuclides were also evaluated. The conclusions from these tests are as follows.

1. Regardless of the variables introduced in the analysis (e.g., water content of the sediment, level of iron additives and carbonation of concrete), diffusion coefficients for both Tc and I varied over a range of about 2 orders of magnitude with Tc being slightly more retarded than I.
2. Degradation of concrete through fracturing and carbonation will not strongly influence diffusion rates primarily because of expected low moisture contents in the waste environment (vadose zone sediments).
3. Within the range of measurements, increasing moisture content of the sediment routinely increased diffusion rates.
4. The addition of metallic iron appeared to have competing effects on diffusion. In some cases, limited reduction of Tc may have occurred. On the other hand, metallic iron appears to have facilitated microcracking during carbonation.
5. In the context of a 1,000 year compliance period, concrete encasement of waste provides a significant delay in radionuclide release into the subsurface.

9. Acknowledgment

This work was conducted at Pacific Northwest National Laboratory, operated by Battelle Memorial Institute for the U.S. Department of Energy under contract DE-AC05-76RL01830. Funding for this project was provided by CH2M Hill Plateau Remediation Company and the Department of Energy Environmental Management Office of Technology Innovation and Development. We would like to thank Steven Baum, Kenton Rod, Sara Rither, John Nelson, and Danielle Saunders for assisting in the sectioning of the concrete soil half-cells. We also thank the anonymous reviewer for helpful comments.

10. References

Abdel Rahman RO, and AA Zaki. 2011. "Comparative study of leaching conceptual models: Cs leaching from different ILW cement based matrices". *Chemical Engineering Journal. doi:10.1016/j.cej.2011.08.038.*

Al-Khayat H, MN Haque, and NI Fattuhi. 2002. "Concrete Carbonation in Arid Climate." *Materials and Structures* 35(7):421-426.

ANSI. 1986. *Measurement of the Leachability of Solidified Low-Level Radioactive Wastes by a Short-Term Test Procedure,* ANSI/ANS-16.1, American Nuclear Society, Chicago.

Brown DA, BE Fulton and RE Phillips. 1964. "Ion Exchange Diffusion: I. A Quick-Freeze Method for the Measurement of Ion Diffusion in Soil and Clay Systems." *Soil Sci. Soc. Am. Proc.* 28:628-632.

Crank J. 1975. *The Mathematics of Diffusion.* Second Edition. Oxford University Press, New York.

EPA—U.S. Environmental Protection Agency. 1999. *Toxicity Characteristics Leaching Procedure. Test Methods for Evaluating Solid Wastes – Physical and Chemical Methods SW-846*. Method 1311.

Finney DJ. 1971. *Probit Analysis; A Statistical Treatment of the Sigmoid Response Curve*. Third edition, Cambridge University Press, New York.

Garrabrants AC, and DS Kosson. 2003. "Modeling Moisture Transport from a Portland Cement-Based Material During Storage in Reactive and Inert Atmospheres." *Drying Technology* 21(5):775-805.

Garrabrants AC, F Sanchez, C Gervais, P Moszkowicz, and DS Kosson. 2002. "The Effect of Storage in an Inert Atmosphere on the Release of Inorganic Constituents during Intermittent Wetting of a Cement-Based Material." *Journal of Hazardous Materials* 91(1-3):159-185.

Garrabrants AC, F Sanchez, and DS Kosson. 2004. "Changes in Constituent Equilibrium Leaching and Pore Water Characteristics of a Portland Cement Mortar as a Result of Carbonation." *Waste Management* 24(1):19-36.

Gervais C, AC Garrabrants, F Sanchez, R Barna, P Moszkowicz, and DS Kosson. 2004. "The Effects of Carbonation and Drying During Intermittent Leaching on the Release of Inorganic Constituents from a Cement-Based Matrix." *Cement and Concrete Research* 34(1):119-131.

Lamar DA. 1989. *Measurement of Nitrate Diffusivity in Hanford Sediments via the Half-Cell Method*. Letter Report to Westinghouse Hanford Company, Pacific Northwest National Laboratory, Richland, Washington.

Mann FM, RJ Puigh II, SH Finfrock, EJ Freeman, R Khaleel, DH Bacon, MP Bergeron, PB McGrail, and SK Wurstner. 2001. *Hanford Immobilized Low-Activity Waste Performance Assessment: 2001 Version*, DOE/ORP-2000-24, Rev. B. Pacific Northwest National Laboratory, Richland, WA.

Sanchez F, C Gervais, AC Garrabrants, R Barna, and DS Kosson. 2002. "Leaching of Inorganic Contaminants from Cement-Based Waste Materials as a Result of Carbonation during Intermittent Wetting." *Waste Management* 22(2):249-260.

Sanchez F, AC Garrabrants and DS Kosson. 2003. "Effects of Intermittent Wetting on Concentration Profiles and Release from a Cement-Based Waste Matrix." *Environmental Engineering Science* 20(2):135-153.

Serne RJ, WJ Martin, VL LeGore, CW Lindenmeier, SB McLaurine, PFC Martin, and RO Lokken. 1989. *Leach Tests on Grouts Made with Actual and Trace Metal-Spiked Synthetic Phosphate/Sulfate Waste*. PNL-7121, Pacific Northwest Laboratory, Richland, WA

Serne RJ, RO Lokken, and LJ Criscenti. 1992a. "Characterization of Grouted LLW to Support Performance Assessment." *Waste Management* 12(2-3):271-287.

Serne, RJ, LL Ames, PF Martin, VL LeGore, CW Lindenmeier, and SJ Phillips. 1992b. *Leach Testing of in Situ Stabilization Grouts Containing Additives to Sequester Contaminants*. PNL-8492, Pacific Northwest Laboratory, Richland, Washington.

Serne RJ, JL Conca, VL LeGore, KJ Cantrell, CW Lindenmeier, JA Campbell, JE Amonette, and MI Wood. 1993. *Solid-Waste Leach Characterization and Contaminant-Sediment Interactions. Volume 1, Patch Leach and Adsorption Tests and Sediment Characterization*. PNL-8889, Pacific Northwest Laboratory, Richland, WA.

Serne RJ, WJ Martin, and VL LeGore. 1995. *Leach Test of Cladding Removal Waste Grout Using Hanford Groundwater*. PNL-10745, Pacific Northwest Laboratory, Richland, WA.

Wellman DM, SV Mattigod, GA Whyatt, L Powers, KE Parker, and MI Wood. 2006. *Diffusion of Iodine and Rhenium in Category 3 Waste Encasement Concrete and Soil Fill Material.* PNNL-16268, Pacific Northwest National Laboratory, Richland, WA.

Wood, MI, R Khaleel, PD Rittman, AH Lu, SH Finfrock, TH DeLorenzo, RJ Serne, and KJ Cantrell. 1995. *Performance Assessment for the Disposal of Low-Level Waste in the 200 West Area Burial Ground.* WHC-EP-0645, Westinghouse Hanford Company, Richland, Washington

4

Environmental Migration of Radionuclides (^{90}Sr, ^{137}Cs, ^{239}Pu) in Accidentally Contaminated Areas of the Southern Urals

V. V. Kostyuchenko, A. V. Akleyev, L. M. Peremyslova,
I. Ya. Popova, N. N. Kazachonok and V. S. Melnikov
Urals Research Center for Radiation Medicine, Chelyabinsk,
Russian Federation

1. Introduction

In the late 1940s, the facility Mayak Production Association (Mayak PA) for weapon grade plutonium production was put into operation in the vicinity of the town of Kyshtym. The technology used in plutonium production involved generation of high-level waste. A number of accidents that occurred at the plant were associated with inadequate radioactive waste storage techniques. In 1949-1956, radioactive waste with total activity of about 1.8×10^{17} Bq (4.9 MCi) was discharged into the Techa River which resulted in contamination of all river system components. Currently, at late time after the beginning of contamination, ^{90}Sr and ^{137}Cs still remain essential dose-forming radionuclides on the Techa River. In 1957, the East-Urals Radioactive Trace (EURT), and in 1967 the Karachai Radioactive Trace (KRT), were formed. A distinguishing feature of the radionuclide composition of the releases on the EURT at late time is the prevalence of ^{90}Sr and a minimum content of ^{137}Cs. The composition of radioactive dust on the 1967-Trace is represented primarily by Cs and Sr isotopes in less accessible biological forms compared to those observed on the EURT (Fig. 1). In contaminated areas, measurements of soil contamination levels, analysis of the patterns of radionuclide migration, changes in their biological accessibility, transfer of radionuclides from soil to vegetation, milk and vegetable produce have been conducted on regular basis. Specific activity of ^{90}Sr measured in cross sections of the river at most of the riverside villages has decreased to permissible values since the start of observations in 1960. In flooded areas of the bank line, the processes of deepening of radionuclides into soil and a more uniform distribution of radionuclide contents over the soil layer at a depth of 1.5 m were observed. Mean content of ^{90}Sr in milk produced in the riverside villages has declined to permissible values. On EURT and KRT, of the total radionuclides contained in the soils, 80% remain deposited in the upper 20-cm layer. Biologically accessible and insoluble forms of ^{90}Sr and insoluble forms of ^{137}Cs are prevalent. Reduction in radionuclide content in milk has taken place over the first 1-2 years due to deepening of radionuclides into soil and a decrease in their biological accessibility. The main factor that caused cleansing of radionuclides from food chains was radioactive decay and reduced biological accessibility of radionuclides in soils.

1.1 Natural-climatic characteristics of the affected territory

As a result of the accidents at the Mayak AP, a number of rivers, water basins and lands of the southern and middle zone of the Trans-Urals region were contaminated. The EURT occupies over 3/4 of the forest and forest-steppe part of the Trans-Urals region where there are numerous lakes, swamps, all kinds of depressions and pits, wood lands and forest outliers which account for non-uniformity of radioactive fallouts. The most common are chernozemic-meadowy and meadowy-chrnozemic soils. The Trans-Urals region has a typical continental climate which is formed by the air masses coming from the Atlantic Ocean. The wind conditions of the region are characterized by prevalence of westerly winds. Whirlwinds are not an infrequent phenomenon. Species of wood prevailing in the forest zone include pines and the main hardwood species – birches and aspens. The floodplain vegetation includes grassy and woody-shrubby species. Birches and willows are encountered in the floodplain. Miscellaneous herbs are characteristic of the Techa floodplain in the middle and lower reaches of the river.

Fig. 1. Schematic map of radiation accidents in the Southern Urals

1.2 Methods of the study

The key contaminants of the environment at late phase of the accident are represented by a small number of radionuclides: ^{90}Sr, ^{137}Cs, ^{239}Pu and a few other. Samples of water, bottom sediments, aquatic vegetation, soil and grass were collected at 8 control cross sections on the Techa River in accordance with conventional standard methods of sample drawing. (Basic Requirements..., 1999, in Russian [Основные требования..., 1999]). Samples of bottom sediments were collected at a depth of 60 cm using metallic tubes, the volume of the samples ranged from 20 to 100 cm³. To allow assessment of the distribution and deposits of radionuclides in floodplain soil, samples were drawn at a depth of up to 150 cm. Samples of grass were taken from the area of 0.25 m². Water was sampled at a depth of 10-50 cm using a sampling device. Measurements of radioactive contamination in the riverside villages and the adjacent areas, contents of radionuclides in food products, migration of activity from soil to food products, horizontal and vertical migration of radionuclides in soils were conducted In the EURT zone. Processing and preparation of food products and other samples to be used in a corresponding assay were conducted using the generally accepted methodology. Measurements of ^{137}Cs and other gamma-emitting radionuclides were performed using the gamma-spectrometric method. ^{90}Sr concentration in samples was measured by the radiochemical separation of daughter ^{90}Y using monoisoochthyl-methyl ether of phosphonic acid (1966) and a subsequent measurement of its activity in a small-background β-metric installation analogous to UMF-2000 based on a flame photometric control of strontium carrier yield. ^{90}Sr measurement error accounts for 20% at activities of <0.7 Bq/g, and for 10% at higher activity levels. The range of measured values was $0.02\text{-}1 \cdot 10^5$ Bq/dm³.

The method for measuring ^{239}Pu involves increasing concentration of plutonium ions, and cleaning of isotopes using anion-exchange tar followed up by electro-chemical precipitation on steel disks. Measurements of α-activity were performed in an α-spectrometric installation. Identification and measurements of specific activity of plutonium isotopes was conducted using an indicator mark (^{236}Pu or ^{242}Pu) with a known activity ranging from 1.5 to $1 \cdot 10^5$ Bq/kg, dm³ preliminarily introduced into the sample.

2. Techa River

2.1 Hydrographic characteristics of the Techa River area

The Techa River basin catchment area is situated between the mountainous Urals region and the Tobol River valley. The water from the Kasli-Irtyash lake system is flowing through the Techa River. After the construction of the Mayak plant was started, a cascade of industrial water reservoirs (IWB) was built for storing low-level liquid waste. In 1956 reservoir B-10, and in 1963-1964 reservoir B-11 were built. Since 1966, the IWRs have been functioning as stagnant water reservoirs. Since 1965, the Techa River was conditionally assumed to rise from the dam of reservoir B-11 (Fig. 2). The drainage of the Kasli-Irtyash lake system takes place through the bypass canals (BC).

From dam P-11 to the village of Muslyumovo, the river flows through a wide valley with numerous swamps. The floodplain is mainly composed of peaty soils. The river bottom is peaty-silty and uneven. Along the river stretch from the village of Muslyumovo and further on, the river flows through a flat even country. The surface of the floodplain is meadowy, formed by sandy-loam soils. The bottom of the river is sandy, miry and loamy at places. In the lower reaches of the river, the width of the valley ranges from 240 m to 2 km. The floodplain is meadowy, loamy, usually flooded during high water. The watercourse is moderately winding. The river is mostly supplied with snow water. The swampy floodplain of the upper reaches retain a considerable amount of thaw water. The river's tributaries are water-short. Floods usually occur in April. Low water lasts till mid-October. Water discharge during the low water periods increases along the river length from 0.84 m³/d near Muslyumovo to 2.62 m³/d near the village Klyuchevskoye. The coefficient of ground water supply accounts for 10-30% of the total river drainage.

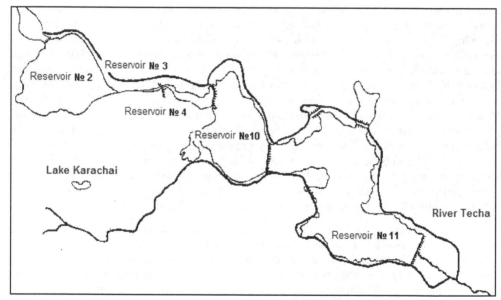

Fig. 2. Schematic map of the Techa cascade of reservoirs

The contamination of the Techa River was caused by the unavailability of reliable technology for reprocessing and storage of liquid radioactive waste (LRW). The approximate total releases into the Techa River over the period 1949-1953 were as follows: ^{89}Sr+^{140}Ba 240 kCi, ^{90}Sr -320 kCi, ^{137}Cs -350 kCi, REE 740 kCi, ^{95}Zr+^{95}Nb 37 kCi, ^{103}Ru+^{106}Ru kCi (Glagolenko Yu.G., 1966).

2.2 Radioactive contamination of water

During the initial period, the studies of radioactive contamination of water were based on measurements of β-emitting nuclide activity. The most well-systematized data were presented in (Marey A.N., 1959). The highest level of β-activity was observed in water in 1951; it was decreasing appreciably with advancing years and increasing distance from the site of releases (Fig. 2). The activity of α-emitters in water was significantly lower. The dependences governing the changes in the concentration of these emitters are similar to those identified for β-emitters. Reduced concentrations of radionuclides in the river water with increasing distance from the release site were accounted for primarily by the dilution processes in the water flow, sedimentation and radionuclide sorption by bottom sediments.

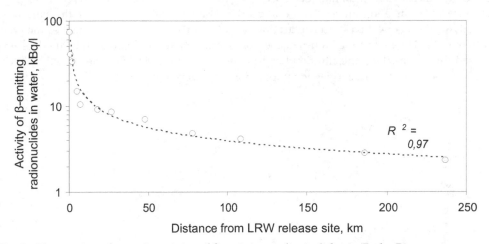

Fig. 3. Changes in volumetric activity of β-emitting radionuclides in Techa River water as a function of distance from LRW release site, 1952 (Marey A.N., 1959).

Nuclide composition of the river water sampled in the middle and lower reaches was for the first time determined in 1951 (Table 1). It was established that a significant proportion of activity of the radionuclides cesium, yttrium, cerium and plutonium is transported by the river stream down the rivercourse on clayey and sandy particles. The same applies to zirconium and niobium. The radionuclides strontium and ruthenium are transferred with river stream mostly in the dissolved state. The basic source of inflow of suspended particles is the surface-slope drainage from the catchment area.

Village name	β –activity in water, Bq/l	^{90}Sr	^{137}Cs	REE
Brodokalmak	3.1·10⁴	54.2	14.0	34.0
Bisserovo	1.7·10⁴	50.0	32.0	19.3
Pershinskoye	2.1·10⁴	55.8	24.0	15.2

Table 1. Radiochemical composition of the river water as of 5.08.1951, %

The results of the researches conducted in 1963 showed that small amounts of radionuclides (from 0.001 to 0.014%) (Yu.G. Mokrov, 2002) were carried by the bottom alluvium to the Techa River.

The construction of the Techa cascade of water reservoirs for storing low-level sewage water and re-directing medium-level waste to Karachai Lake resulted in reduced concentrations of radionuclides in river water and bottom sediments. By that time, the radionuclides ^{90}Sr and ^{137}Cs became the most important contaminants of the Techa River. The long-term dynamics of radionuclide content measured in river water (e.g., at Muslyumovo) up to 1990 was characterized by persistent reduction in ^{90}Sr and ^{137}Cs concentration. Instability and periodical increases in radionuclide concentrations have been observed in river water (Fig. 4) since 1994. In addition to that, ^{90}Sr concentration is stably exceeding the currently permissible level of 4.9 Bq·l^1. Concentrations of ^{137}Cs in river water are less stable along the watercourse (Fig. 5), but the values of the volumetric activity of this radionuclide does not exceed the permissible concentration for drinking water (11 Bq·l^{-1}), the role played by ^{137}Cs in radiation exposure of the riverside population is not very significant.

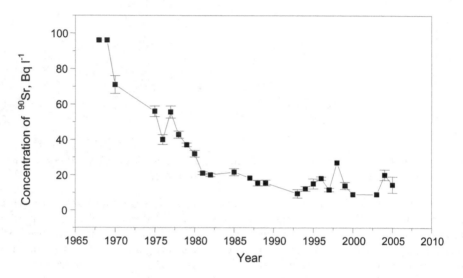

Fig. 4. Specific activity of ^{90}Sr in water of the river stretch vis-à-vis Muslyumovo.

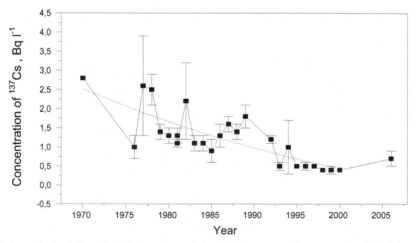

Fig. 5. Specific activity of ^{137}Cs in water of the river cross-section opposite Muslyumovo.

Data on contents of plutonium isotopes in Techa River water are scarce. Measurements of specific activity of ^{238,239,240}Pu in river water was initiated by the URCRM researchers in 1993

(Table 2). The main source of radioactive contamination of river water is the Techa cascade of reservoirs. Additional contamination is accounted for by desorption of radionuclides from the contaminated floodplain and the river bottom sediments.

In 2009, specific activity of tritium was for the first time determined in water of the Techa river (Fig. 6). Presented in the figure are concentrations of ^{90}Sr and tritium in water of the Techa River over its total length down to its confluence with the Isset River measured in samples taken within a week's time in August 2009.

Sampling site	Distance from dam 11, km	Years		
		1994	2000	2002
Assanov Bridge	4,5	3.2 ± 0.4 (3)	0.8 (2)	0.4 (2)
Chelyabinsk-Yekaterinburg Bridge	14	3.1 ± 0.8 (3)	0.46 ± 0.2 (4)	1.5 (2)
Muslyumovo	40	3.1 ± 1 (3)	0.14 (2)	-
Brodokalmak	71	-	-	-
Russkaya Techa	95	0.3 (2)	-	-
Nizhnepetropav-lovskoye	103	-	0.6 ± 0.2 (3)	-
Lobanovo	119	1.6 (2)	0.36±0.2 (3)	-
Verkhnyaya	141	2.5 (1)	1.0 (2)	-
Pershinskoye	170	2.7 (1)	-	-
Zatechanskoye	195	2.1 (2)	0.3 ± 0.1 (4)	-

Note: numbers of samples are given in brackets

Table 2. Specific activity of $^{238, 239, 240}$Pu in Techa River water in 1994-2002, Bq·m^{-3}

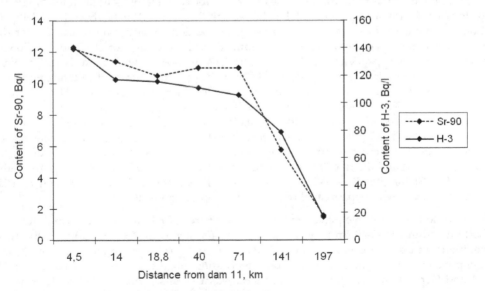

Fig. 6. Content of ^{90}Sr and ^{3}H in water of the Techa River

Concentrations of ^{90}Sr in water measured over the river length has changed from 12 to 1.6 Bq/l, tritium activity ranged from 140 to 17 Bq/l, which is 4-fold higher than the global level. The mean value of ^{3}H/^{90}Sr ratio in river water is 10.9±1.2. The total estimated carry-over of ^{90}Sr with the Techa river run-off into the Isset River over the period from 1958 through 2001 amounted 2.2×1014 Bq (Yu.A. Izrael, 2000).

Zone	Volumetric activity (Bq l^{-1})								
	^{90}Sr			^{3}H			^{137}Cs		
	Range of values	Mean	Median	Range of values	Mean	Median	Range of values	Mean	Median
TCR	983-5149	3007	3162	763-5935	2881	3251	1.9-914.5	102.3	32.9
Techa River	3.8-32.9	14.6	11.8	31-263.2	129	116	0.028-2.4	0.39	0.17
Drained area	0.05-173.9	27	14.1	6-15.8	10.2	9.5	0.05-1.51	0.55	0.49
Tributaries	0.15-0.18	0.15	0.15	9.5-9.8	9.7	9.5	-	-	-

Table 3. Summarized statistical data on volumetric activities of ^{90}Sr, ^{3}H, ^{137}Cs in water bodies of different origin

Table 3 shows summarized statistical data on volumetric activities of the radionuclides measured in the Techa River, TCR, surface and ground waters of the catchment area in the upper reaches of the river for the period from 2008 through 2010. In our subsequent calculations, the median values of ^{90}Sr and ^{3}H were used for TCR water, as well as the ^{3}H/^{90}Sr ratio equal to 1, median values of ^{90}Sr and ^{3}H for Techa River water, ^{3}H/^{90}Sr radioactivity ratio equal to 10, rounded median values of volumetric activity of ^{3}H equal to 10 Bq l^{-1} in samples of surface and soil water taken in the catchment area of the upper reaches of the river. Concentrations of radionuclides in water of the tributaries corresponded to the background values obtained for contents of radionuclides in surface waterways in the area influenced by the Mayak PA: ^{90}Sr - 0.15 Bq l^{-1}, ^{3}H - 9.7 Bq l^{-1}. Concentrations of ^{137}Cs were found to be below the detectability level of activity for the volumes sampled. The comparisons of volumetric activity values for ^{90}Sr and ^{3}H allowed tracing a direct dependence of the concentrations in river as well as in TCR. The only mechanism determining concentrations of ^{90}Sr in water is the process of intermix of waters with different initial radionuclide concentrations in different proportions. Also, uranium isotopes (^{234}U, ^{238}U) were used as a radionuclide of reference in relation to ^{90}Sr (^{234}U, ^{238}U). Uranium was selected because, unlike ^{3}H, its isotopes represent ions dissolved in water, as is ^{90}Sr (Mokrov Yu.G., 2000); and also because under oxidizing conditions characteristic of surface waters, uranium existing in the form of uranyl-ion (UO_2^{+2}) is weakly sorbed by floodplain soils and the river's bottom sediments. The difference between the values of the ratio ^{3}H/^{90}Sr for river and TCR waters (10 and 1) is accounted for by dilution of TCR effluent seepage with bypass canal waters in the proportion 1 to 10.

In order to assess the role played by the catchment area of the upper reaches and the river bottom sediments (0-40 km) in contamination of water with ^{90}Sr, we applied the two-component mixing model ($X_M=X_A \times f + X_B \times (1-f)$, where X_M is the end mixture, X_A and X_B are components, and f is the compound coefficient) using mixture parameters obtained for ^{3}H. ^{3}H and ^{90}Sr volumetric activities were measured at cross-sections located at 3.5 km and 40 km from dam 11, respectively. The values of ^{90}Sr volumetric activities for waters flowing into the river from the catchment area of the upper reaches range from 4.3 Bq L^{-1} to 19.3 Bq l^{-1}, the average value amounting to 9.24 Bq l^{-1}. It was concluded based on the calculations that entry of 70% of the total activity of ^{90}Sr into the Techa watercourse results from TCR effluent seepage drained through the bypass canal system. The proportion of ^{90}Sr activity contributed by washing out of radionuclides from the floodplain and by desorption from bottom sediments accounts for 30%.

2.3 Contamination of bottom sediments with ^{90}Sr and ^{137}Cs

According to the data of the first investigations, the highest concentrations of radionuclides in bottom sediments were observed in the reaches close to the release site and in the area of Assanov swamps (30 km from the release site). The lowest level of contamination was registered over the last 40 km stretch down to the outfall. A large amount of activity was accumulated in the surface layer (Table 4) (Glagolenko Yu.G., 1966).

Distance from the release site, kn	Depth, cm		
	0-1	5-7	10-12
Release site	103045-486550	703-77071	262-62678
7	50098	8621	703
18	47101	16539	10730
33	6734	4625	37
48	1961	1184	59
78	2035	292	44
109	888	540	373
138	1221	355	222
186	255	215	133
223	307	252	-

Table 4. Specific β-activity of bottom sediments, August 1952, kBq·kg^{-1} (Glagolenko Yu.G., 1966).

Distance from the dam 11, km	^{137}Cs	Σ Pu
1	2	3
1.8	27491	61
2.2	14874	33
3.8	45103	133
5.8	23199	27
6.7	7104	24
13.0	15392	57
14.0	17020	60
23.0	32190	78
41.0	666	4
44.0	8325	11

Table 5. Distribution of contamination density in bottom sediments along the length of the river in 1991-1992, kBq·m^{-2}

It was established that radioactivity is best of all accumulated by silt and clay material, while sandy soils manifest a lower rate of accumulation. It was demonstrated that ^{137}Cs and plutonium isotopes were intensively sorbed by all varieties of soils. [90]Sr, too, is actively sorbed by soils, but it easily enters into exchange reactions which determines its high migration potential (Saurov M.M., 1968). [90]Sr and ^{137}Cs contamination densities in bottom sediments in the early 1990s are shown in Table 5. During spring floods, silt sediments contaminate the surface of the floodplain which maintains the high levels of contamination of the flooded riverside valley. The levels of silt and water radionuclide contamination are interdependent over the river course. The contents of radionuclides in silt and in water, from dam 11 up to the river outfall are steadily decreasing (Figure 7). Specific activity of ^{137}Cs in silts is about 5-fold higher than in water, that of ^{90}Sr if 3-fold higher. Compared to ^{90}Sr, ^{137}Cs contamination densities for silts are about 2-fold higher over the total length of the river.

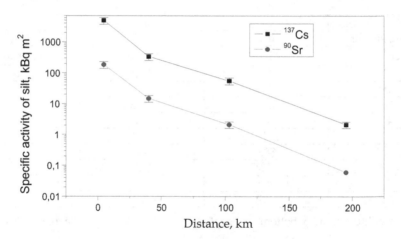

Fig. 7. [90]Sr and ^{137}Cs contamination densities measured in Techa River silts at different distances from dam 11.

Four decades after termination of intensive discharges of radioactive waste into the river, radionuclides deposited in sandy and silty soils migrated to the depth of over 35 cm. The results of the vertical distribution of the radionuclides of interest in the upper reaches of the river are presented in Figures 8 and 9. Compared to ^{137}Cs, the distribution of ^{90}Sr in the bottom soil profiles is more uniform. Maximum values of contamination densities for these radionuclides are in general observed in 0-10 cm layers of soil. [137]Cs is characterized by a more dramatic decline of contamination density values in lower layers of soil (at the depth of 20-35 cm). In the upper reaches, additional inflow of radionuclides due seepage from TCR and washout of radioactivity from the lands adjacent to the river is observed; in the mid-stream area the contribution of desorption processes from the upper layer is larger.

With distance from the release site, the proportion of exchangeable and mobile forms of ^{90}Sr is increasing, on the contrary, the proportion of poorly-accessible forms of ^{90}Sr is decreasing.

2.4 Contamination of the floodplain with ^{90} Sr and ^{137}Cs

Spring overflows of the Techa, and particularly the flood of 1951 contributed to intensive radioactive contamination of the riverside area. The studies of the contents of radioactive substances in floodplain soils started in 1951. The results obtained allowed an insight into the patterns and intensity of the riverside contamination (A.N. Marey, 1959). The width of the floodplain where radioactive contamination was detected did not usually exceed 150-200 m (Table 6). In the upper reaches, in the Assanov swamps area, the overflow reaches 3000 m. A consequent decrease in the levels of floodplain contamination with increasing distance from the release site, and a decrease in the ^{137}Cs/^{90}Sr ratio should also be noted (Table 7).

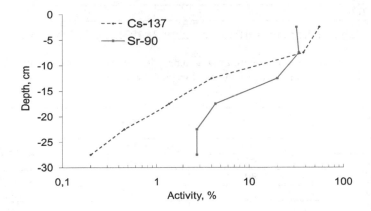

Fig. 8. Distribution of radionuclides along the depth of sandy bottom sediments in the upper reaches of the river.

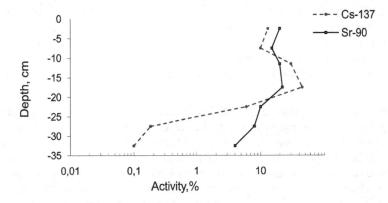

Fig. 9. Distribution of radionuclides along the depth of silty-clayey bottom sediments in the upper reaches of the river.

Distance from the release site, km	Bank	End of the inundable floodplain
Release site	7548	141
18	2442	18
33	407	22
48	74	15
78	481	4
105	200	60
109	63	33
128	33	4
138	4	23
212	3	22

Table 6. Specific β-activity in the surface layer in 1952 - 1954, kBq·kg^{-1} (Marey A.N., 1959)

Changes in the width of the riverside area is, as a rule, determined by soil relief; maximum concentrations of radionuclides are registered in lowland areas.

Distance from dam11, km	^{137}Cs	^{90}Sr	^{137}Cs /^{90}Sr
67	210	150	1.4
102	95	60	1.5
135	80	80	1.0
152	35	100	0.3
237	5	42	0.1

Table 7. Changes in contents of ^{137}Cs and ^{90}Sr in floodplain soils along the river course, Ci·km^{-2} (Marey A.N., 1959)

The first maximum level of floodplain contamination density is observed in the area of Assanov swamps: from dam 11 to the distance of 7.5 km. The second peak takes place in lower reaches, at the distance of 37 km from the dam in the area of Muslyumovo swamps. In these areas, the incessant winter run-off of radioactive substances with TCR waters is accumulated. Most thoroughly the floodplain soils are washed during high waters. Surface-downslope waters wash upper layers of the soil, flood waters wash upper layers of the flooded river bank, and ground waters wash deeper layers of the floodplain soils.

With distance along the watercourse, the levels of floodplain contamination with ^{137}Cs are appreciably decreasing. In the swamps of the upper reaches, the values of contamination with the radionuclides of interest amount to 150-550 Ci·km^{-2}, the values for the middle reaches are 20-30 Ci km^{-2}, and the respective value for the area close to the estuary is 5 Ci·km^{-2}. The dynamics of reduction in ^{90}Sr contamination density values assumes a more monotonous character. The results of measurements of ^{137}Cs and ^{90}Sr contamination densities conducted by us in the floodplain in 2005 are presented in figure 10. It can be seen that the level of floodplain contamination with ^{137}Cs is higher than the respective value obtained for ^{90}Sr, with the exclusion of the last locality (village Zatechanskoye).

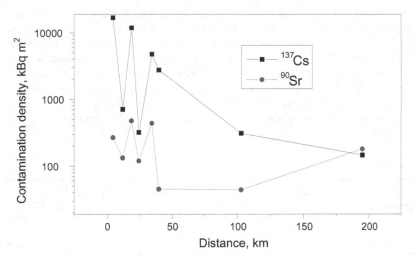

Fig. 10. ^{137}Cs and ^{90}Sr contamination densities in the floodplain along the watercourse, kBq m^{-2}

The study of ^{137}Cs distribution by layers of the floodplain along the watercourse (Table 8) has demonstrated that the upper 20 cm layer is contaminated to the highest extent. However, the measurements conducted in the village Zatechenskoye where sandy soils predominate, the concentrations of radionuclides in the lower and surface layers are actually the same. As has been shown by measurements at most of the sampling sites, the upper 0-5 cm soil layer contains a smaller amount of ^{137}Cs as compared to the underlying layer. This fact indicates that there has been no intensive contamination of the floodplain with ^{137}Cs in the recent years. An extreme non-uniformity in the distribution of ^{90}Sr by soil profiles (Table 9) in different reaches of the river can be accounted for by different physical-chemical and morphological properties of the floodplain soils.

Sampling site	Distance from dam 11, km	Position of layers, cm			
		0-5 см	5-20 см	20-40 см	40-60 см
Assanov bridge	4.5	1438.7	1011.0	1595.3	1059
Maloye Taskino	12	230.7	61.1	7.9	6.9
Nadyrov bridge	18.8	593.2	10084.4	1281.3	439.0
6 km below Nadyrov bridge	24.5	129.6	28.4	53.7	44.9
Buslyumovo swamp	34.7	660	62.1	784.3	354
Muslyumovo	40	277.1	95.4	17.7	10.9
Nizhne-Petropavlovskoye	103	8.1	65.4	19.3	6.4
Zatechenskoye	195	11.3	99.8	31	101.7

Table 8. Distribution of ^{137}Cs by floodplain soil layers over the length of the river, kBq m^{-2}

Most of the floodplain soils (Assanov Bridge, Nadyrov Bridge) are characterized by depletion of the upper 0-10 cm layer, presence of maximum ^{90}Sr concentrations at the depth of 10-30 cm with a dramatic drop in concentrations measured deeper along the profile. This is caused by washing of the upper layer with the surface waters. In the locality of Muslyumovo, the distribution of ^{90}Sr by soil layers is influenced by the continuous run-off from the overlying swampy layers. Further along the watercourse, the processes of cleaning radionuclides from the upper soil layers and a more uniform distribution along the depth of the soil become prevalent. (Nizhnepetropavlovskoye and Zatechenskoye).

Sampling site	Distance from dam 11, km	Position of layers, cm			
		0-5 см	5-20 см	20-40 см	40-60 см
Assanov bridge	4,5	20	78,5	112,6	49,4
Maloye Taskino	12	51,8	143,8	12,6	2,1
Nadyrov bridge	18,8	27,2	394,1	136,9	37,6
6 km below Nadyrov bridge	24,5	46,2	39,2	26,6	8,3
Buslyumovo swamp	34,7	27,7	72,1	233,9	103,7
Muslyumovo	40	10,7	33,8	6,5	2,1
Nizhne-Petropavlovskoye	103	2,5	16,4	3,6	0,74
Zatechenskoye	195	26,3	239,8	321,1	24,5

Table 9. Distribution of ^{90}Sr by floodplain layers along the watercourse, kBq·m⁻²

From 1991 through 2005, the total content (reserve) of ^{137}Cs at the sampling site (Fig. 11) decreased from 574.5 MBq·m⁻² to 52.8 MBq·m⁻², accumulating mainly at the depth of 30 cm, and lower, at 40 cm above the aquifer. The total content of ^{90}Sr in soil samples taken in the floodplain (Fig. 12) decreased from 14.4 to 8.8 kBq·m⁻². Over this period, equalization of the contamination was noted at the depth of 55 cm, while in the lower layer an increase in contamination level was observed, with radionuclides moving all the way to the aquiferous layer, obviously due to intensive seasonal washouts of the bog soils.

In 2009, samples were taken in the floodplain area close to Assanov swamps at the depth of 80 cm within 30, 70, 106 and 250 m of the shoreline. Below, an aquifer composed of blue clay was situated. It was shown that ^{137}Cs and ^{90}Sr contamination densities in the floodplain soil (figure 13) were declining depending on distance from the shoreline. The reserve of ^{137}Cs exceeds that of ^{90}Sr at any distance, and their ratio changed at different distances from 22-2 times. The distance from 30 to 70 m is characterized by the lowest levels of contamination due to wash-off by ground waters.

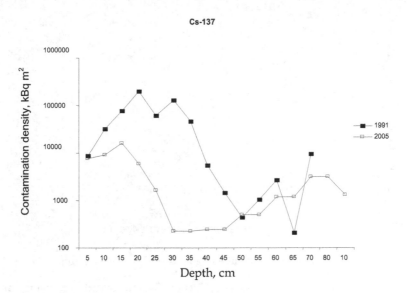

Fig. 11. Contamination densities for ^{137}Cs in the Techa floodplain area at sampling site "Assanov swamps" within 6.5 km of dam 11, 10 m from the shoreline.

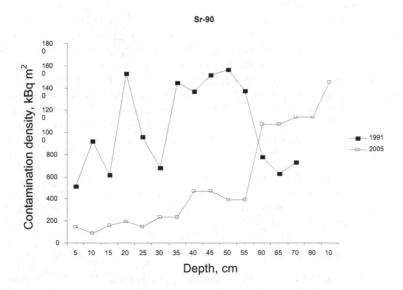

Fig. 12. ^{90}Sr contamination densities in the Techa River floodplain at 6.5 km from dam 11, left bank, within 10 of the shoreline.

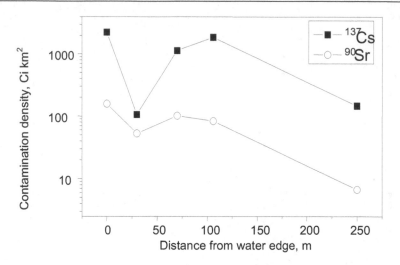

Fig. 13. ^{137}Cs and ^{90}Sr contamination densities for soils of Assanov swamps, Ci km^{-2}

In samples of floodplain taken in the area of Assanov swamps at different depths and sites, prevalent are water-soluble forms of radionuclides deposited in deep layers (70-80 cm) of riverside soil (up to 26%) in localities where they migrate in the direction of the watercourse. A slight increase in exchangeable and mobile forms of ^{137}Cs in the lower layer was observed.

In 2005 and 2006, samples of ground and surface waters were taken in the area of Assanov and Muslyumovo swamps with the aim to determine the actual scope of influence exerted by deposited radionuclides on concentrations of radioactivity in swamp water. During cold seasons, when migration processes are at their minimum, concentration of ^{90}Sr in flow channels is high, reaching 187 Bq l^{-1}, the concentration level registered in ground water is 14-19 Bq l^{-1}. In spring, in the portion of the swamp area, where there is no distinctly outlined water flow, ^{90}Sr concentrations in water amount to 80-100 Bq l^{-1} in localities far removed from the river banks. In flow channels with a sufficiently dynamic water flow, radionuclide concentrations reach about 40 Bq l^{-1}. In Muslyumovo swamps, water sampled in dead channels contains 11-25 Bq l^{-1} of ^{90}Sr with radionuclide concentration level in river water of 6.2 Bq l^{-1}. The data obtained point out to the fact that it is mainly ^{90}Sr which is leached from the floodplain to water, the same applies to ^{137}Cs, but to lesser extent.

Thus, it was demonstrated that the floodplain and bottom deposits are the key source of secondary contamination of the river water with radionuclides. The level of radioactive contamination of these river components is, in its turn, determined by the radioactive run-offs from the Techa cascade of reservoirs.

3. East-Urals Radioactive Trace

3.1 Radioactive contamination of soil, vegetation, food products with ^{90}Sr and ^{137}Cs in the early years after the 1957 accident

A distinguishing feature of the radioactive emission on the EURT is the presence in it of all basic uranium fission products and a minimum content of the ^{137}Cs (Table 10).

Radionuclide	%	Scope of release, PBq
^{89}Sr	traces	-
^{90}Sr+^{90}Y	5.4	2.0
^{95}Zr+^{95}Nb	24.8	18.4
^{106}Ru+^{106}Rh	3.7	2.7
^{137}Cs	0.36	0.26
^{144}Ce+^{144}Pr	66.0	48.7
^{147}Pm, ^{155}Eu	traces	-
^{239}Pu	traces	0,0014

Table 10. Characterization of the radioactive releases and the initial reserve of radionuclides on the EURT outside the Mayak PA industrial site (Avramenko V.I. et al., 1977)

The settling of the radioactive mixture from the cloud which was wind-drawn in the north-eastern direction from the explosion site resulted in the formation of the East-Urals Radioactive Trace (EURT). The Trace encompassed parts of Chelyabinsk, Sverdlovsk and Tyumen oblasts. The length of the EURT is 300 km, its width is 30-50 km. According to refined data, (Korsakov Yu.D. et al., 1996), in 1957 the area with contamination density (η) for ^{90}Sr >2 Ci km^{-2} was 560 km^2, that with η>12 Ci km $^{-2}$ was 230 km^2, with η>50 Ci·km^{-2} – 120 km^2, with η>200 Ci·km^{-2} –50 km^2, with η>800 Ci·km^{-2} – 16 km^2, and with η> 2 000 Ci·km^{-2} – 8 km^2. There were over 200 populated localities in the Trace area, including several towns and industrial communities. At early time after the accident, the population of the EURT area was exposed to radiation-related hazards, including, in the first place, external exposure to γ-radiation due to prevalence of γ-emitting nuclides in the deposited mixture and, in the second place, internal exposure resulting from intakes of radionuclides contained in food products produced in the localities. As the activity of γ-emitting nuclides (which decayed almost completely 6-7 years after the accident) decreased, the radiation hazards were mostly determined by the radionuclide ^{90}Sr.

The first measurements conducted on the contaminated territory showed that the γ-radiation dose rate was proportional to the distance from the accident site. According to measurements conducted during the first year after the accident in the area with contamination density amounting to 1 Ci km^{-2} for^{90}Sr, the dose in air due to γ-radiation was 1R (G.N. Romanov, 1963). Direct exposure to β-radiation only occurred in areas with contamination density of over 1 500 Ci·km^{-2} for ^{90}Sr. According to data presented in (A.Ya. Kogotkov, 1968), a decrease in relative contents of ^{90}Sr in the composition of the mixture depending on distance from the accident site, and, respectively, an increase in the contents of ^{144}Ce and ^{137}Cs. Soil contamination densities in some localities may differ by an order of magnitude, or greater.

The coniferous woods were affected most heavily. At the distance of 12.5 km from the contamination source, a total loss of pine woods was registered in the summer of 1958. Mass loss of birch forests were only observed in areas with contamination densities of over 4 000 Ci·km^{-2}. Migratory birds were only affected in the spring of 1958, after dose rates in the tree crowns had decreased 10-fold. A reduced number of bird's nests was noted in the areas with contamination level of over 2 000 Ci·km^{-2} for ^{90}Sr. No loss of animals was registered. In the ensuing years, due to the fact that the contaminated area was made into a sanitary-protection zone, the number of hares, roes and elks increased considerably.

During the first days after the accident, radioactive contamination of grass estimated relative to 1 Ci/km^2 of soil contamination was $1.5 \cdot 10^6$ decay min^{-1}·kg^{-1}. Radioactive contamination of individual food products was very high (Table 11). Since cattle and other domestic animals were fed contaminated forage, contamination of milk and meat was of structural rather than superficial nature. It should be noted that contamination of milk was registered as early as the first 2-3 days after the accident, and that of meat on days 10-12 (R.M. Alexakhin et al., 2001).

During the first days after the accident, radioactive contamination of grass estimated relative to 1 Ci/km^2 of soil contamination was $1.5 \cdot 10^6$ decay min^{-1}·kg^{-1}. Radioactive contamination of individual food products was very high (Table 11). Since cattle and other domestic animals were fed contaminated forage, contamination of milk and meat was of structural rather than superficial nature. It should be noted that contamination of milk was registered as early as in the first 2-3 days after the accident, and that of meat on days 10-12 (R.M. Alexakhin et al., 2001).

Locality	Soil	Water from water sources	Grass	Milk	Meat	
					muscles	bones
1	2	3	4	5	6	7
Berdyanish	21016	52	360380	115	58	92
Saltykova	33760	104	97162	218	-	-
Galikayeva	2405	104	34262	-	6	74
Kasli	18	-	32	-	-	-
Russ. Karabolka	4810	42	22200	-	6	137
Yugo-Konevo	192	0.02	15207	-	1.2	10
Yushkovo	3	-	-	3.1	3	71
Boyevskoye	266	-	755	-	-	-
Bagaryak	74	-	8399	4.4	-	-
Kamensk-Uralsky	-	0.004	-	-	2.8	1.5
Pozarikha	-	-	440	-	-	-

Note: the results include ^{40}K activity

Table 11. β-activity in food products and environmental entities in the EURT area as of 20.10.1957, kBq/kg, l

Food products were the main contributor to radiation exposure of the population. The main cause of the reduction in contents of radionuclides in food products over time is the reduction of radionuclide contents in soils. Table 12 shows the basic composition of radionuclides observed in food products and the dynamics of cleaning the food from radionuclides. Among those radionuclides, ^{144}Ce+^{144}Pr prevailed during the first 3 years, later on ^{90}Sr gained priority.

Beginning from the spring-summer season in 1958, an additional contamination of vegetation and agricultural plants occurred due to a downwind migration of radionuclides from areas with a higher level of contamination density. The proportion of surface contamination of grain crops accounted for 10-15% (P.P. Lyarsky, 1962).

Contamination levels measured for food products of the first post-accident harvest reaped in the fields situated in the Trace zone with contamination densities of 0.2-0.5 Ci·km^{-2} for ^{90}Sr or higher, were higher than the permissible limit legally valid at that time: 1300 decay min^{-1}·kg^{-1} (22 Bq·kg^{-1}).

Time after the accident, years	Key radionuclides, %					Other
	^{90}Sr+^{90}Y	^{137}Cs	^{144}Ce+^{144}Pr	^{95}Zr+^{95}Nb	^{106}Ru+^{106}Rh	
1 (1957)	10.1	1.5	76.9	10.1	1.3	
2 (1958)	12.9	2.8	82.4	0.2	1.5	
3 (1959)	7.0	8.0	80.6	-	4.2	<1%
5 (1961)	90.3	2.7	6.2	-	0.6	
10 (1966)	98.8	1.2	-	-	-	

Table 12. Composition of radionuclide observed in food products at different time after the accident

3.2 Current levels of soil contamination

The studies conducted in 2006-2009 in the villages Allaki, Bagaryak, Bulzi, Tartar Karabolka and Yushkovo located around the perimeter of the EURT of sanitary-protection zone. The soils of the pasture lands adjacent to these villages are mainly dark-grey clay and leached chernozem.

The mean dose-rate value of gamma-radiation in the localities of Karabolka, Musakayeva and Bagaryak is 0.12 µSv·hr^{-1} which is comparable to natural background values for Chelyabinsk oblast. The mean pasture land contamination density for Sr^{90} in the localities of Tatarskaya Karabolka and Musakayeva is 5.9 kBq·m^{-2}, that for Bagaryak is 2-2.5 times higher. Pasture land contamination density for ^{137}C is 12.9-24.8 kBq·m^{-2}. Mean contamination densities measured in soils of kitchen gardens attached to houses in Bagaryak was 77.5 kBq·m^{-2}. Mean specific activity in grass measured in different areas along the EURT axis ranged from 11 to 15 thousand Bq·kg^{-1} for Sr^{90} and from 3 to 60 Bq·kg^{-1} for ^{137}C. Forms of Sr^{90} and ^{137}C encountered in soils sampled in the frontal portion of the Trace (close to Alabuga Lake) have been identified (Table 13). Attention is drawn to the increased number of exchangeable forms of Sr^{90} and mobile forms of ^{137}C. During the early years after the fallout, the total amount of Sr^{90} was only encountered in soluble state. A decade later, the proportion of exchangeable forms accounted for 65-75%, and the value has not changed since then.

According to data presented by V.V. Martyushov et al. (1996) 36 years after the contamination, the content of exchangeable forms of [137]C and plutonium did not exceed 3% and 1%, respectively. The proportion of poorly accessible forms of [137]C and plutonium reached 95-98%. The content of water-soluble forms of [137]C and plutonium accounts for less than 1%. Water-soluble forms of Sr^{90} are mostly found in cationic compounds (72-76%).

Radionuclide	Physical-chemical forms of radionuclides , %			
	Water-soluble	Soluble in 1H CH_3COONH_4 (exchangeable)	Soluble in 1H HCl (mobile)	Solid residue (poorly accessible)
[90]Sr	2.8 ± 0.5	73.2 ± 0.4	16.3 ± 0.6	7.7 ± 0.4
[137]Cs	2.7 ± 0.4	3.1 ± 1.0	19.4 ± 1.5	74.8 ± 2.5

Table 13. Forms of radionuclides identified in the upper 5-cm layer of soil in the vicinity of Alabuga Lake

In 2009, soil samples were taken at 16 sampling sites at a distance of 20 km from the Mayak PA perpendicular to the EURT axis. Five of those sites are situated in a birch forest. The total contamination density in the soils of the rhizogenic layer and forest litter was, on the average, equal to 737 kBq/m² for ^{90}Sr, and 41.2 kBqκ/m² for [137]Cs. It should be noted that 10.5% of the contamination density for ^{90}Sr and 14.1% of contamination density for ^{137}Cs are contributed by the debris layer. At 10 points situated in the hayfield extending from the forest up to Alabuga Lake ^{90}Sr contamination density ranged from 161 to 350 kBq/m², ^{137}Cs contamination density measured in the hayfield ranged from 34 to 93 kBq/m². Contamination densities measured in the wood were higher than those measured in the hayfield.

At a distance of 30 km from the Mayak PA in an area along the Trace axis samples of soils were taken at a depth of 10 cm. [90]Sr contamination density was found to range from 2.2 kBq/m² to 55.9 kBq/m², that for ^{137}Cs ranged from 2.2 to 50.7 kBq/m². At the periphery of the Trace, contamination density for ^{137}Cs exceeded that for [90]Sr 2-5-fold. The contribution made by the debris layer varies significantly. In afforested and steppified areas with well developed steppe debris layer, the contribution of the debris layer to the contamination density for ^{90}Sr reaches 37%, for [137]Cs – 8.8%.

At a distance of about 40 km from the Mayk PA at a right angle to the Trace axis, ^{90}Sr and ^{137}Cs contamination density distribution was determined for the ploughed layer (0-20 cm). As of the date of the measurements, there were only arable lands that had no sod cover were available for measurements. The highest measured contamination density was 162 kBq/m² for ^{90}Sr and 24.4 kBq/m² for [137]Cs

Samples of sod-podzol soil were taken on the Trace axis close to Bolshoi Irtysh Lake (about 55 km from the Mayak PA). Contamination densities in the upper 0-20 cm layer and in the ground litter was 308 kBq/m² for ^{90}Sr, and 20 Bq/m² for [137]Cs. It is noteworthy that 11.2% of ^{90}Sr and 18.0% of ^{137}Cs are contributed by the ground litter. The total contamination densities in soil (0-20 cm) and ground litter along the Trace axis are presented in Figure 14.

The studies performed have shown that soil contamination densities close to the EURT axis are still high, even at the present time; besides, in a number of cases a considerable portion of radionuclides is contained in forest debris layer or steppe litter. Contamination levels mostly depend on the distance from the contamination source, i.e. from the Mayak PA. The highest levels of radioactive contamination were identified in the 0-5 cm layer of soil. Below the root-inhabited layer, 2.7%-57.2% of ^{90}Sr and 28.4%-41.1% of ^{137}Cs are deposited. In all types of soil, ^{90}Sr is encountered in biologically accessible forms, and, ^{137}Cs is contained in poorly-soluble compounds.

3.3 Vertical migration of ^{90}Sr and ^{137}Cs through soil profiles

To allow assessment of soil contamination levels, two main parameters are used, viz., specific activity of the radionuclide in soil (Bq·kg^{-1}) and contamination density (Bq·m^{-2}) which takes into account the total contamination density in all n soil layers. There is no direct relationship between these two parameters. It was established that over the time period since the 1957 accident the radionuclides had migrated to a significant depth. Actually, samples taken at all the sampling sites showed that the highest level of contamination with ^{137}Cs and ^{90}Sr was detected in the upper level of soil and in debris layer (in meadow soils it was found in sod cover and in thick felt of the steppe). The ratio of radionuclide specific activity in the 0-10 cm layer to activity in the 10-20 cm layer did not depend on the summarized contamination density. The value of this ratio is mostly influenced by the type of the ecosystem: the mean value of this ratio for ^{137}Cs in forest ecosystems is 20.4±4.4, and in meadows it is 2.9±1.6. For ^{90}Sr, the differences are insignificant: 3.4±0.9 in forests, and 2.8±1.5 in meadows. The ratio of specific activity of ^{137}Cs in debris cover to activity in the 0-10 cm soil layer amounted on an average to 0.5±0.1, and for ^{90}Sr to 1.5±0.2. Although the specific activity of ^{137}Cs and ^{90}Sr in the litter layer is sufficiently high, it does not exert substantial influence on the summarized contamination density since the volume weight of the litter layer is by two orders of magnitude lower than the volume weight of soil.

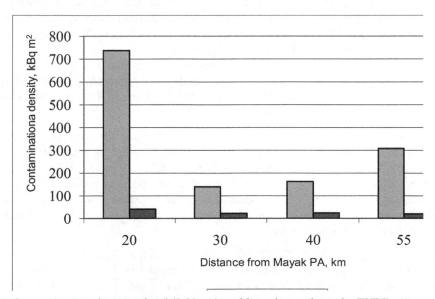

Fig. 14. Contamination density of soil (0-20 cm) and litter layer along the EURT axis as a function of distance from the source of contamination.

Distribution of ^{90}Sr and ^{137}Cs in soil profiles was studied in 2008-2010 for 3 types of soils: gray forest, sod-podzol and chernozem. Gray forest soils are most prevalent in the forest-steppe zone. They are formed in leaf woods and mixed woods of the Trans-Urals region. Distribution of ^{90}Sr and ^{137}Cs in gray forest soil profiles was determined at a sampling site within 20 km of the Mayak PA. The highest specific activity of ^{90}Sr (1.6×10^4 Bq·kg^{-1}) and ^{137}Cs (1.3×10^3 Bq·kg^{-1}) was measured in the lower layer of the forest litter which contains a half-decayed tree waste. ^{90}Sr activity in the upper layer of the forest litter is actually equal to activity found in the 0-5 cm soil layer, ^{137}Cs activity is 2.7 times lower. However, the volume weight of the forest litter is low that is why contamination density summarized for its upper and lower layers is lower than that in the underlying 0-5 cm soil layer where the main amounts of ^{90}Sr (60%) and ^{137}Cs (50%) are deposited. The tree waste of 2008 sampled on October 31 demonstrated an increase in contamination density of the tree waste by 2.4 kBq·kg·m^{-2} for ^{90}Sr which accounts for 2.1% of the tree waste contamination measured in spring of 2008. Over the 50 years since the accident, no shift of the maximum values with increasing depth through the soil profiles has taken place. At the same time, both ^{90}Sr, and ^{137}Cs, though in small amounts, have at least, reached the depth of 170-175 cm. It should be added that ^{137}Cs which is considered to be less capable for vertical migration, is evidently migrating more actively than ^{90}Sr, 2.7% of ^{90}Sr, and 28.4% of ^{137}Cs are deposited at a depth of over 20 cm.

Sod-podzol soils which most often are found in the northern part of the Trace, are formed under coniferous woods. The distribution of ^{90}Sr and ^{137}Cs along the profile of sod-podzol soils is comparable to the distribution observed for gray forest soils. The highest specific activity of ^{90}Sr (1.0×10^4 Bq·kg^{-1}) and ^{137}Cs (1.2×10^3 Bq·kg^{-1}) was found in the lower part of the forest waste, the highest contamination density for ^{90}Sr (39.6%) and ^{137}Cs (40.4%) was measured in the upper 0-5 cm soil layer. The forest waste and soil layer up to 20 cm in depth contain 96.2% of ^{90}Sr and 58.9% of ^{137}Cs. At the depth of over 20 cm 3.7% of ^{90}Sr and 41.1% of ^{137}Cs are deposited below 20 cm. As can be seen from the comparison of the distributions, the mobility of ^{90}Sr in the sod-podzol soils differs insignificantly from that registered in gray forest soils, the mobility manifested by ^{137}Cs is significantly higher. Already at the depth of 25-30 cm, the specific activities of ^{90}Sr and ^{137}Cs actually differ one from the other, however, in lower layers ^{137}Cs takes the first position.

There occur in steppefied areas of the EURT weak northern or leached chernozems which have been ploughed up for a long time period. It should have been expected, therefore, that the distribution of radionuclides through the 0-20 cm plough-layer would be more uniform and speedy. The site for taking soil samples from the chernozem profile is located close to the EURT axis, however, the contamination density for ^{90}Sr turned out to be low, viz., 17.8 kBq·m^{-2} for ^{90}Sr and 40.2 kBq·m^{-2} for ^{137}Cs over the whole profile. ^{137}Cs was found to be significantly prevalent in each layer.

The highest specific activity for ^{90}Sr (92.2 Bq·kg^{-1}) and ^{137}Cs (161 Bq·kg^{-1}) was observed in the debris cover which was characteristic of other soil types too. The highest contamination density was measured in the 0-5 cm layer for ^{90}Sr (10.4%) and in the 5-10 cm layer (10.9%), and for ^{137}Cs in the 0-5 layer (21.4%). At the depth of over 20 cm, 57.2% of ^{90}Sr, and 37. 4% of ^{137}Cs were deposited.

The analysis of ^{90}Sr distribution patterns in the EURT soils were made using data of researches conducted in 1963-2008 (figures 15-17). It can be seen that the distribution of ^{90}Sr in the profile of the 30 cm layer is well described by the following exponential function: $y=ae^{-bx}$, where y is the content of radionuclides calculated as percentage of the total contamination density in the 30-cm layer, x is reference number of sample taken in the 5-cm soil layer. Using the coefficient b it becomes possible to calculate the depth at which a decrease in contamination density to a preset level takes place. ^{90}Sr which settled on the soil surface is slowly migrating to deeper layers, and the coefficient b is decreasing (Table 14).

In gray forest soils, the value of coefficient b correlates with the number of years that have passed since the accident ($r = -0.94$, $p = 0.02$). In sod-podzol soils, especially intensive ^{90}Sr migration is going on.

So far, no shift in maximum ^{90}Sr and ^{137}Cs activities down the soil profile has been observed in any soil types of interest (gray forest, sod-podzol, chernozem). The highest specific activity of both ^{90}Sr, and ^{137}Cs in natural and fallow lands is retained in the lower layers of forest litter and or steppe debris. High specific activity of ^{90}Sr and ^{137}Cs is also retained in 0-5 soil layer. In deeper soil layers, the activity of these radionuclides is decreasing rapidly and reaches the minimal values in the 25-40 cm layers of eluvial horizons. In illuvial horizons, radionuclide activity is slightly decreased. In general, the specific activity of ^{90}Sr and ^{137}Cs in natural soils, beginning at a depth 20-25 cm is relatively stable. The calculations have shown that the 180-300 cm soil layer contains about 28% of ^{137}Cs and 18% of ^{90}Sr of the total contamination density in soil layer 0-300.

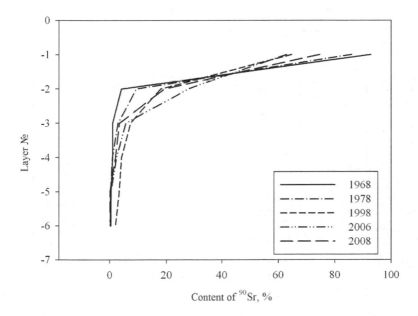

Fig. 15. Dynamics of ^{90}Sr distribution in 30-cm layers of gray forest soil

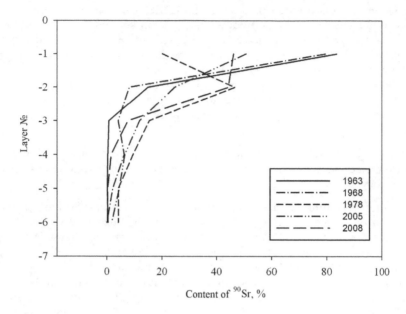

Fig. 16. Dynamics of ^{90}Sr distribution in 30-cm layers of sod-podzol soil

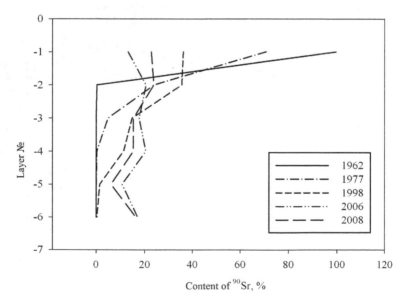

Fig. 17. Dynamics of ^{90}Sr distribution in 30-cm layers of chernozem

Year of sampling	b	R2
Gray forest		
1968	3.07	0.999
1978	2.15	0.999
1998	1.17	0.993
2006	0.96	0.991
2008	1.38	0.999
Sod-podzol		
1963	1.76	0.993
1968	2.17	0.989
1978	0.30	0.435
2005	0.70	0.998
2008	0.62	0.831
Chernozem		
1962	5.78	1.000
1977	1.16	0.997
1998	0.45	0.874
2006	0.01	0.008
2008	0.16	0.555

Table 14. Parameters of equations describing ^{90}Sr distribution in the profile of the 30-cm soil layer

Most of the researchers engaged in studies of radionuclide distribution in soil have noted that in the first years after atmospheric fallouts the highest amounts of ^{90}Sr and ^{137}Cs settled in sod cover or forest litter. It is known, however, that vegetable waste of herbaceous type decays within one season, while leaf wood waste decays within 3-4 years. It takes longer for waste of coniferous forests to decay, however, over the 50 years since formation of the EURT the forest litter contaminated by atmospheric fallouts should have decayed long ago, while the steppe cover of chernozem did not develop until 1991. Thus, the high specific activity currently observed in the forest litter is due not only to the initial fallouts in 1957 and 1967, but also to a high rate of local radionuclide turnover and continuous fallouts of activity.

The reserve of radionuclides in the vegetation mat and upper soil layers is constantly replenished due to vegetation waste. The waste cover is, in its turn, contaminated due to atmospheric fallouts and transfer of radionuclides to the top by the root system. Since actually the total surface biomass of vegetation is transferred to the waste cover at the end of the year, the yearly carry-over of ^{90}Sr to the surface is determined by the annual productivity of the ecosystem.

Waste of woody vegetation accounts for just a portion of the yearly increment, in addition to leaves and grass, the waste composition includes slowly decaying wood, needles and strobiles. That is why radionuclides are deposited not only in soil but also in wood and litter cover. Mineralization of decayed biomass involves an increase in the relative contents of radionuclides in the waste litter. Specific activity of ^{90}Sr in the ground litter was 2.2±0.7 times higher than in grass at a distance of 20 km from the contamination source, and 6.4±36 times higher at a distance of 30 km; the respective values for ^{137}Cs were 14±8 and 12±8. Table

15 presents specific activity values for radionuclides deposited in litter cover and the upper layer of soil. ^{90}Sr activity measured in the upper layer of the litter cover is actually the same as that in the upper layer of the soil, and activity in the lower layer is substantially higher. The difference between levels of ^{137}Cs activity in the upper and lower layers of the litter cover is even more substantial than that found for ^{90}Sr. Ratio of ^{90}Sr specific activity to the activity in the 0-10 cm soil layer at a distance of 20 km from the Mayak PA is 0.7±0.3, and that at 30 km is 2.1±1.0. The ratio for ^{137}Cs is 0.08±0.04 and 0.4±0.4. Since contamination levels of grass are higher at longer distances, it can be assumed that currently the uptake of ^{90}Sr and ^{137}Cs by the biomass is going on through the root system. ^{90}Sr activity in leaves which contribute the largest portion of the vegetation mat is actually the same as that identified in the 0-5 soil layer. However, as was mentioned above, the weight of the vegetation waste per 1 m^2 was not large. On the EURT axis, within 20 km of the Mayak PA, samples of 2008-vegetation waste were taken in a birch forest. Specific activity of ^{90}Sr measured in the waste was 5904 Bq·kg^{-1}, that of ^{137}Cs 54 Bq·kg^{-1}. The fall of the aboveground phytomass to the forest floor contributed 2.3 kBq·kg m^{-2} of ^{90}Sr and 028 kBq·kg m^{-2} of ^{137}Cs. This accounted for 2.1% and 3.4% of contamination density in forest litter for ^{137} and ^{90}Sr, respectively, and for 0.36% and 0.8% of the total contamination density in the 0-20 cm layer of forest litter and soil for ^{90}Sr and ^{137}Cs, respectively.

Distance from Mayak PA, km	Soil	Upper layer of litter cover		Lower layer of litter cover		0-5 cm soil layer	
		^{90}Sr	^{137}Cs	^{90}Sr	^{137}Cs	^{90}Sr	^{137}Cs
20	gray forest	7.1×10³	0.18×10³	16.3×10³	1.3×10³	7.8×10³	0.48×10³
	chernozem	0.6×10³	0.02×10³	1.2×10³	0.2×10³	1.1×10³	0.3×10³
30	chernozem	29	5	92	161	31	143
55	Sod-podzol	2.8×10³	0.06×10³	10×10³	1.2×10³	1.9×10³	0.2×10³

Table 15. Soil and litter contamination levels, Bq ·kg^{-1}

Therefore, the plant litter vegetation waste contributes an insignificant proportion of contamination density of the plant litter and upper layer of soil.

It was found out as a result of measurements of the proportions of radionuclides occurring in different forms and having different degrees of accessibility to plants that ^{90}Sr is mostly encountered in exchangeable form (64-85%), a larger amount of exchangeable strontium being deposited in sod-podsol soils than in gray forest and chernozem. The largest amount of ^{137}Cs is strongly bound together and it occurs in acid-soluble form (4-36%) or as a solid residue (27-82%). In all cases, the content of ^{137}Cs accessible to plants in the 5-10 layer proved to be higher, and the content of ^{137}Cs in the form of solid residue was significantly lower.

Thus, as of today, the studies of all the soil types have shown that the major part of ^{90}Sr and ^{137}Cs activity is deposited in the upper layer.

3.6 Dependencies governing migration of ^{90}Sr and ^{137}Cs in the soil-grass chain and forest products

In 2005-2006, samples of grass and soils taken in the most heavily contaminated areas of the EURT were processed, following which proportionality factors were assessed (PF). Specific activity of ^{90}Sr in grass ranged from 70 to10940 Bq·kg^{-1}, that of ^{137}Cs from 10 to 997 Bq·kg^{-1}. Proportionality factor (Bq/kg in grass/Bq/kg in soil) was within the range 0.2-2.1, the mean value of transfer factor calculated as a ratio of specific activity of radionuclides in grass to soil contamination density, Bq·kg^{-1}/ Bq·kg^{-2}, was 15.9±8.5 for ^{90}Sr, while for ^{137}Cs it was lower: 3.1±3.4. The difference in transfer factor values may depend on soil type and diversity of plant species composing the cover layer. It should be noted that transfer factor values are slightly higher for sod-podzol soils in the EURT zone.

Transfer factor estimated for grass in the EURT zone has changed insignificantly over the recent 10 years, it is 10.3 Bq·kg^{-1} / Bq·kg^{-2} for ^{90}Sr. The mean transfer factor calculated in 1997 for ^{90}Sr in pasture soils in the EURT zone was 13.5 Bq·kg^{-1}/Bq·kg^{-2}. Transfer factor values calculated for ^{90}Sr relative to standard soil contamination of 1 Ci·km^{-2} (37 kBq·m^{-2}) amounted to 230-270 Bq·kg^{-1} / 37 Bq·kg^{-2}.

Contamination with ^{90}Sr of all sampled wild-growing berries exceeds the permissible limit 3.5-13.5 times, ^{90}Sr proportionality factor is higher than that determined for vegetables and grain (Table 16). Specific activity of ^{137}Cs in grass and berries does not exceed the permissible limit. It should be taken into consideration that wild-growing berries in the EURT zone present the highest hazard compared to other food products in view of the contribution they can make to dietary intakes of ^{90}Sr.

Sampling site	Species	Specific activity of fresh berries,Bq·kg^{-1}		proportionality factor, (Bq·kg^{-1} in berries)/ (Bq·kg^{-1} in soil)	
		^{90}Sr	^{137}Cs	^{90}Sr	^{137}Cs
20 km from Mayak PA	Wild strawberry	813	5.2	0.140	0.0165
	stone bramble	243	2.4	0.042	0.0076
30 km from Mayak PA	Wild strawberry	213	1.0	0.254	0.0054

Table 16. Contamination of wild-growing berries with ^{90}Sr and ^{137}Cs

Table 17 presents data on levels of contamination of fresh mushrooms. In 2008, specific activity of ^{90}Sr and ^{137}Cs in all samples was found to be significantly lower than the permissible limits (50 Bq·kg^{-1} for ^{90}Sr and 500 Bq·kg^{-1} for ^{137}Cs). Proportionality factors and transfer factors were significantly lower than those obtained for grass, vegetables and grains sampled in 2007. It can be assumed that mushrooms growing in the EURT zone present no hazards for the population as their contribution to internal dose is insignificant.

Sampling site	Species	Specific activity of fresh mushrooms, Bq·kg⁻¹		Accumulation factor, (Bq·kg⁻¹ in the fruit body)/(Bq·kg⁻¹ in soil)		Proportionality factor, (Bq·kg⁻¹ in fruit body) / (Bq·kg⁻² in soil)	
		^{90}Sr	^{137}Cs	^{90}Sr	^{137}Cs	^{90}Sr	^{137}Cs
50 km from Mayak PA	Birch mushroom	0.47	0.58	0.0039	0.0113	0.0328	0.0941
	Coral milky cap	1.2	0.97	0.0101	0.0189	0.0839	0.1573
	Mix	0.7	0.9	0.0059	0.0175	0.0489	0.1460
	Milk mushroom yellow	0.34	4.6	0.0041	0.0176	0.0305	0.1447
	Blewits	0.59	19.9	0.0071	0.0762	0.0529	0.6259
	Arachnoid muchroom, excellent	0.63	18.5	0.0076	0.0709	0.0565	0.5819
	Mix	1.4	8.2	0.0169	0.0314	0.1255	0.2579
	Blewits white	0.7	3.9	0.0092	0.0169	0.0749	0.1383
	Mix	1.7	18.8	0.0531	0.0450	0.2798	0.3705
20 km from Mayak PA	Honey agaric summer mushroom	9.8	39.0	0.0017	0.124	0.0133	0.9476
30 km from Mayak PA	Yellow boletus	5.5	10.9	0.0065	0.0586	0.0396	0.4950
	Mix of the lamellar	8.9	2.0	0.0106	0.0108	0.0641	0.0908

Table 17. Contamination of mushrooms with ^{90}Sr and ^{137}Cs

3.7 Migration of radionuclides along food chains

Levels and dynamics of ^{90}Sr specific activity in milk produced in villages at the periphery of the EURT in 1958-2006 are presented in figure 24. A dramatic decrease in contents of ^{90}Sr in milk had occurred before 1963 while later on a slow decrease was observed overtime. Since 1960 the period of a 2-fold decrease in ^{90}Sr contents in milk ($T_{1/2 \ eff.}$) was 23 years. The content of ^{137}Cs in milk produced in the villages studied was 2 times lower as compared to that of ^{90}Sr. Currently, this value is1.1 ±0.4 Bq/l for ^{137}Cs.

In 1960, the transfer factor in the soil-milk chain was 340 µBq·l⁻¹/Bq·m⁻², in 2006 it was 45 µBq·l⁻¹/Bq·m⁻². A reduction in the transfer factor resulted from a decrease in biological accessibility of the radionuclide in the soil-pasture grass chain.

Potatoes take the second position after milk in terms of ^{90}Sr contributed by food produce. The dynamics of reduction in specific activity of ^{90}Sr in potatoes has been the sane over the total period of observations, viz., from 7.0 to 0.9 Bq·kg^{-1}, the respective value for ^{137}Cs is from 1.5 to 0.7 Bq·kg^{-1}. Within a number of years, stable values of transfer factor have been established for agricultural products (Table 18).

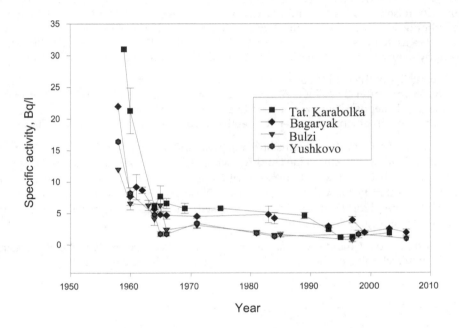

Fig. 24. Dynamics of specific activity of ^{90}Sr in milk produced in EURT villages in the period 1958-2006

Crops	Products	Proportionality factors	
		^{90}Sr	^{137}Cs
Peas	Grain	11.4±6.6	15.2±5.2
Wheat	Grain	14.4±9.1	6.6±1.9
Oats	Grain	5.7±2.5	9.9±4.6
Barley	Grain	6.5±2.1	4.4±1.3
Cabbage	Head of cabbage	4.7±1.6	3.0±1.4
Carrots	Root crop	5.2±1.1	2.2±0.8
Beet	Root crop	43.3±44.7	16.5±15.7
Potato	Tubers	4.8±1.3	2.3±0.5
Corn	Paste (dry)	1.4±0.6	3.6±0.5
Sudanese grass	Paste (dry)	9.0±7.2	46.0±36.7

Table 18. Mean values of proportionality factors for ^{137}Cs and ^{90}Sr measured in the portion of cash crop grown in 0-20 cm layer of gray forest soils

Values of proportionality factor decrease overtime and, as a result, it becomes possible to grow agricultural product with admissible level of conatimation in soils with a higher contaminated level.

4. Karachai Radioactive Trace

In 1997 the area covered by the Karachai Radioactive Trace (KRT) delineated by ^{90}Sr contamination isoline of 7.4 kBq/m^2 (0.2 Ci/km^2) amounted to 1660 km^2. The total amount of the radionuclides deposited on this territory was 800 Ci. Radinuclide composition of dust fallouts was as follows: 32 % of ^{90}Sr+^{90}Y, 47% of ^{137}Cs, 21% of ^{144}Ce+^{144}Pr. Biological accessibility to plants was 90 % for ^{90}Sr, and 12% for ^{137}Cs (Yu.D. Korsakov et al., 1996).

In 1967 about 97% of the total ^{137}Cs deposited in pasture soils settled in the upper 0-3 cm layer. Currently, 38 years after the fallout, ^{137}Cs is accumulated in the soil layer at a depth of 13 cm, and ^{90}Sr is mostly deposited in the soil layer to a depth of 0-20 cm (89.5%). A small portion of radionuclides which settled in the layer at a depth of 0-70 cm migrated to a depth of 70 cm.

In 1967, the transfer factor for contents of ^{90}Sr in grass and soil was 0.09 Bq·kg^{-1} / kBq·m^{-2}. A sharp decline in transfer factor values for ^{137}Cs within the soil-grass chain occurred one year later after it had been cleaned of surface contamination, and in the subsequent years no changes in transfer factor values were noted. During early years, the uptakes of the radionuclide were going on from sod cover and soil, and transfer factor values fluctuated between 0.0025 and 0.005 Bq/kg : Bq/m^2.

Long-term studies of ^{90}Sr and ^{137}Cs transfer from soil to grass were conducted in the grazing land of Sarykulmak village where dairy cattle was grazing (Figure 25).

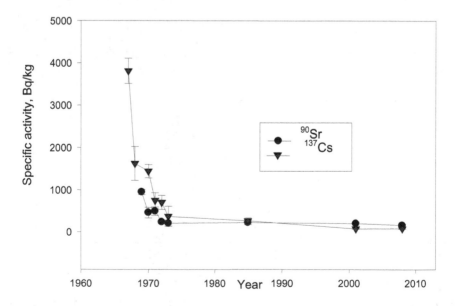

Fig. 25. Levels and dynamics of decrease in radionuclide contents measured in pasture grass

A reduction in specific activity of ^{137}Cs in the pasture grass over the period from 1967 through 2009 is described by the two exponential dependences: $T_{1/2}$ was 1.8 years in the initial period after deposition of radionuclides, and in the subsequent period it was 3.8 years. The first exponent describes the cleaning of grass from surface contamination, the second describes the reduction in uptakes of radionuclides by grass from soil. Over the period from 1967 through 2009, the specific activity of ^{137}Cs in grass decreased 100-fold , that of ^{90}Sr 10-fold.

The initial specific activity of ^{90}Sr in milk late in April, 1967, was 140 Bq/l, that of ^{137}Cs - 237 Bq/l. From 1967 through 2009, the average values of ^{90}Sr and ^{137}Cs specific activity in milk decreased 30-fold. The decrease in ^{90}Sr and ^{137}Cs levels in milk was determined based on the decrease in grass contamination. Beginning from 1970, the period of half-cleaning of milk from ^{90}Sr was 20 years, and from ^{137}Cs 10 years. Proportionality factors for ^{90}Sr in the soil-milk chain estimated on the Karachai Radioactive Trace were on the average 5-fold lower than the values estimated for the EURT for the same time periods elapsed after the radioactive fallouts on the soil. Evidently, the long period for which ^{90}Sr and ^{137}Cs remained deposited in silt and soil on the shores of Karachai led to a reduction in biological accessibility of radionuclides to plants and, as a result, to a reduced uptake of radioactivity by milk. Specific activities of ^{90}Sr in milk are more correctly described by the log-normal distribution. After the accident, the contents of ^{137}Cs in potatoes was insignificant, and overtime it decreased about 10-fold, while the contents of ^{90}Sr decreased 3-fold.

5. Conclusion

1. Major radiation accidents that took place in the Southern Urals during the period 1949-1967 brought about contamination of vast territories with radionuclides and exposure of the local population. The situation on the Techa River involved contamination of river water, bottom sediments and the floodplain with ^{137}Cs and ^{90}Sr. On the East-Urals Radioactive Trace ^{90}Sr was the prevalent contaminant, while on the Karachai Trace biologically poorly-accessible compounds of ^{137}Cs and ^{90}Sr prevailed.

2. The key mechanisms by which contamination of the environment can be eliminated include radioactive decay, reduction in biological accessibility and deepening of radioactive substances in soil. Pronounced sorption capabilities and poor solubility hamper migration of ^{137}Cs in the environment, while ^{90}Sr is more mobile.

3. Due to a number of protection measures implemented from 1965 through 2004 on the Techa River, concentrations of ^{90}Sr and ^{137}Cs in river water have decreased, and currently the specific activity of ^{90}Sr amounts to 10-15 Bq·l^{-1}, and that of ^{137}Cs to 0.5-1.5 Bq·l^{-1}. Radionuclides deposited in bottom sediments migrated to the depth of over 35 cm. Contamination density in the 0-10 cm layer depends on the concentration of radionuclides in the watercourse. ^{90}Sr and ^{137}Cs are revealed in floodplain soil at a depth of over 70 cm where a decrease and averaging of concentrations overtime has been observed. At the present time, it is expedient to retain the constraints on the use of river water.

    After the use of river water was banned in 1956, the major pathway for contribution of radioactivity to dietary intakes has been made by milk (87-95 %) and vegetables. Since 1967, the content of ^{90}Sr in milk, with rare exception, has not exceeded the permissible limit (25 Bq·l^{-1}).

4.   The rate of vertical migration of ^{90}Sr in EURT soils ranges from 0.25 to 0.35 cm year^{-1}, the largest amount of radionuclides which remains deposited in the upper part of the soil profile (0-20 cm) is decreasing with increasing depth. Nonmobile ^{137}Cs is mosty retained in the upper 10-cm layer. Biological accessibility of ^{90}Sr has decreased over the past period 7-10 times and it has not actually changed over the recent years. Currently, 40 years after the accident, the content of fixed forms of ^{90}Sr in soil has reached 34%, that of ^{137}Cs and plutonium 95-98%.

The period of half-cleaning of milk from ^{90}Sr was 2-3 years during the first years after the accident, and during the subsequent period it amounted to 15 years.

5.   On the Karachai Trace, due to the prevalence of ^{137}Cs and ^{90}Sr with poor biological accessibility, the reduction in contamination of soils, grass and food products was going on more speedily (than that registered on the EURT). Specific activity of ^{137}Cs and ^{90}Sr in milk exceeded the permissible limits only during the first month after the accident.

6.   The prognosis for the further development of the radiation situation on the Techa River is determined predominantly by the radioactive runoff from the Techa cascade of reservoirs. With the lapse of time, the part of the EURT territory where the use of agricultural lands is restricted, will diminish. There are no such restrictions in the Karachai Trace area.

6. References

M.I. Avramenko, A.N. Averin., Yu.V. Glagolenko, et al.. 1957 Accident. Assessment of the parameters of the explosion and analysis of the character of radioactive contamination of the territory // Radiation Safety Issues. – 1997. - № 3. - P. 18-28.

R.M. Aleksakhin, L.A. Buldakov, V.A. Gubanov et al. Major radiation accidents: consequences and protection measures. Under the editoship of L.A. Ilyn and V.A. Gubanov / - M., IzdAT, 2001 – 752 p.

Yu.G. Glagolenko, E.E. Dzekun, E.G. Drozhko, G.M. Medvedev, S.I. Rovny, A.P. Suslov. Strategy for radioactive waste management at the Mayak PA// Radiation Safety Issues. №2.-1966.-p 3-10.

Yu.A. Izrael, E.M. Artemov, V.N. Vasilenko, I.N. Nazarov, A.I. Nakhutin, A.A. Uspin, A.M.Kamkin. Radioactive contamination of the Urals region by the Mayak Production Association // Radioactivity in case of nuclear explosions and accidents. (Proceedings of the International Conference, Moscow, April 24-26, 2000, C.-P., 2000.- P. 411-424.

A.Ya. Kogotkov. Behavior of radionuclides in soils of the Mid-Transuralian region: Doctoral thesis (Biological Sciences). – M., 1968. – 556 p.

Yu.D. Korsakov, E.A. Fedorov, G.N. Romanov, L.I. Panteleyev. Assessment of radiation conditions on the territory contaminated due to downwind transfer of radioactive aerosols in the area around the facility in 1967: Abstract // Radiation Safety Issues. - 1996. - № 4. - P. 50-59.

P.P. Lyarsky. Sanitary consequences of contamination of the territory with long-lived products of uranium fission and arrangement for sanitary-prophylactic measures: Doctoral Thesis (Medical sciences). - M., 1962. – 928 p.

A.N. Marey. Sanitary consequences of discharges of radioactive waste from atomic facility into water basins: Doctoral thesis (madical sciences) / Biophysics Instituteof the Ministry of Health, USSR. - M. - 1959. - 441 p.

V.V. Martyushov, D.A. Spirin, G.N. Romanov, V.V. Bazylev. Dynamics of the state and migration of Strontium-90 in soils of the East-Urals Radioactive Trace // Radiation Safety Issues. -1996.-№3.-p. 28-35.

Yu.G. Mokrov. Reconstruction and prognosis of radioactive contamiation of the Techa River. Part I. Role of weighted particles in the processes of radioactive contamination of the Techa Rivert in 1949- 1951 – Ozyorsk: Editorial-Publishing Center VRB, 2002.- 176 p.

Basic requirements of samplig (MU 2.6.1.715-98 and MUK 2.6.1.016-99).

M.M. Saurov. Radiation-hygienic assessment of the natural migration of the population chronically exposed to uranium fission products: Doctoral thesis (Madical sciences).-M.,1968.-663 p.

Compedium of methods for measurements of radioactivity in the environment. Methods for radiochemical analysis. Editor: G.A. Sereda, Z.S. Shulenko. GidrometeoIzdat, M., 1966, 51 p.

Consequences of anthropogenic radiation exposure and rehabilitation issues of the Urals region. Editor: S.K. Shoigu. M.,2002, 287 p.

Nature of Chelyabinsk Oblast. – Chelyabinsk, 1964.

V.P. Yulanov. Pine growth peculiarities (Pinus silvestris) in areas contaminated with radioactive substances: Candidate,s thesis (Biological sciences). - Chelyabinsk, 1965. – 154 p.

6.1 Original references in Russian language

Авраменко М.И., Аверин А.Н., Глаголенко Ю.В. и др. Авария 1957 г. Оценка параметров взрыва и анализ характеристик радиационного загрязнения территории // Вопросы радиационной безопасности. – 1997. - № 3. - С. 18-28.

Алексахин Р.М., Булдаков Л.А., Губанов В.А. и др. Крупные радиационные аварии: последствия и защитные меры. Под общей редакцией Л.А.Ильина и В.А.Губанова/ - М., ИздАТ, 2001 – 752 с.

Глаголенко Ю.Г., Дзекун Е.Г., Дрожко Е.Г., Медведев Г.М., Ровный С.И., Суслов А.П. Стратегия обращения с радиоактивными отходами на ПО «Маяк»//Вопросы радиационной безопасности №2.-1966.-с 3-10.

Израэль Ю.А., Артемов Е.М., Василенко В.Н., Назаров И.Н., Нахутин А.И., Успин А.А., Кямкин А.М. Радиоактивное загрязнение Уральского региона производственным объединением "Маяк" // Радиоактивность при ядерных взрывах и авариях. (Труды Международной конференции, Москва, 24-26 апреля 2000, С.-П., 2000.- С. 411-424.

Коготков А.Я. Поведение радионуклидов в почвах среднего Зауралья: Дис. ...докт. биол. наук. – М., 1968. – 556 с.

Корсаков Ю.Д., Федоров Е.А., Романов Г.Н., Пантелеев Л.И. Оценка радиационной обстановки на территории, загрязненной в результате ветрового переноса радиоактивных аэрозолей в районе предприятия в 1967 г.: Реферат // Вопр. радиационной безопасности. - 1996. - № 4. - С. 50-59.

Лярский П.П. Санитарные последствия загрязнения территории долгоживущими продуктами деления и организация на ней санитарно-профилактических мероприятий: Дис...докт. мед. наук. - М., 1962. – 928 с.

Марей А.Н. Санитарные последствия удаления в водоемы радиоактивных отходов предприятия атомной промышленности: Дисс. на соискание степени докт. мед. наук. / Институт биофизики МЗ СССР. - М. - 1959. - 441 с.

Мартюшов В.В., Спирин Д.А., Романов Г.Н., Базылев В.В. Динамика состояния и миграции стронция-90 в почвах Восточно-Уральского радиоактивного следа // Вопросы радиационной безопасности. -1996.-№3.-с. 28-35.

Мокров Ю.Г. Реконструкция и прогноз радиоактивного загрязнения реки Теча. Часть I. Роль взвешенных частиц в процессе формирования радиоактивного загрязнения реки Теча в 1949- 1951 гг. – Озерск: Редакционно-издательский центр ВРБ, 2002.- 176 с.

Основные требования, лежащие в основе отбора (МУ 2.6.1.715-98 и МУК 2.6.1.016-99).

Сауров М.М. Радиационно-гигиеническая оценка естественного движения населения, подвергшегося хроническому воздействию продуктов деления урана: Дис. ...докт. Мед.наук.-М.,1968.-663 с.

Сборник методик по определению радиоактивности окружающей среды. Методики радиохимического анализа. Под редакцией Середы, Г. А., Шулепко, З. С. Гидрометеоиздат, М., 1966, 51 с.

Последствия техногенного радиационного воздействия и проблемы реабилитации Уральского региона. Под.ред. С.К.Шойгу. М.,2002, 287 с.

Природа Челябинской области. - Челябинск, 1964.

Юланов В.П. Особенности роста сосны (Pinus silvestris) на территории, загрязненной радиоактивными веществами: Дис. ...канд.биол.наук. - Челябинск, 1965. - 154с.

5

Modelling Groundwater Contamination Above High-Level Nuclear-Waste Repositories in Salt, Granitoid and Clay

Michal O. Schwartz
MathGeol
Germany

1. Introduction

Modelling the groundwater contamination in the aquifer above a repository for high-level nuclear waste is a complex task. The usual procedure is based on modelling the transport of radionuclides in one or two dimensions only (NAGRA, 2002; Kosakowski, 2004; Keesmann et al., 2006; SKB, 2006; Nykyri et al., 2008). However, modern computing techniques allow to perform simulations in three dimensions, which are a more realistic representation of the geosphere or far field. This is shown for candidate repositories in salt (Gorleben, Germany) and granitoid (Olkiluoto, Finland). To date, is not possible to conduct similar three-dimensional simulations for a repository in clay because the necessary input data are not available. Nevertheless, it can be shown that the results obtained from the simulations with a granitoid-hosted repository are useful for evaluating the performance of a clay-hosted repository. The objective of this comparative study is to provide tools that facilitate the identification of disposal sites and enhance the transparency of decision-making processes.

2. Computer codes for flow-transport models

The generation of a gas phase is an important issue for a repository in salt. Therefore, both one-phase (liquid only) and two-phase (liquid and gas) simulations are performed for the salt-hosted Gorleben repository. The two-phase simulations are performed with the TOUGHREACT code (Xu et al., 2005). This is a numerical simulation program that calculates chemically reactive non-isothermal flows of multi-phase fluids and geochemical transport in porous and fractured media. The program was developed by introducing reactive chemistry into the multi-phase flow code TOUGH2 (Pruess et al., 1999). The code performs a fully coupled simultaneous solution of mass and energy balance for multicomponent fluids, combined with a multiphase extension of Darcy's Law and Fickian diffusion. The program consists of a number of functional units with flexible and transparent interfaces. The governing equations are discretised using integral finite difference for space and fully implicit first-order finite difference in time. The resulting non-linear algebraic equations are solved by Newton-Raphson iteration. The simulations are performed with the ECO2 module, which treats the aqueous phase as a mixture of water and salt. The dependence of density, viscosity, enthalpy and vapour pressure on salt concentration is taken into account, as well as the effects of salinity on gas solubility in the liquid phase and related heat of solution.

The one-phase scenario for Gorleben is calculated with TOUGH2-MP (Zhang et al., 2008), the parallel version of the TOUGH2 code. The EOS7R module is used, which calculates the transport of a parent radionuclide and a daughter radionuclide. The aqueous phase is treated as a mixture of water and brine.

For a granitoid-hosted repository, the potential gas release is negligibly small. Therefore, only a one-phase scenario is calculated for the Olkiluoto repository. The TOUGHREACT code with the ECO2 module is used for these simulations.

Fig. 1. Location of the candidate repositories Gorleben (Germany), Olkiluoto (Finland) and Opalinus Clay (Switzerland)

3. The salt-hosted repository at Gorleben

3.1 Repository layout

The repository (Figs. 1 and 2) will accommodate 5,440 waste containers with a total of 8,550 t U of burnt fuel (Keesmann et al., 2005). The waste containers, which are made of Mn-Ni steel, have a length of 5.5 m and a diameter of 1.6 m (Javeri, 2006). They will be emplaced in horizontal disposal drifts, which will be connected to two or more shafts by a system of access drifts. The envisaged disposal depth is about 800 m. There are 260-280 m Quaternary-Tertiary siliciclastic sediments above the present mine workings within the Zechstein salt dome.

3.2 Near field

The near field is the hydrochemical regime of the backfilled underground workings in the salt dome and the far field is hydrochemical regime of the caprock and cover sediments. The far-field simulations use the near-field releases as input. To date, only two near-field scenarios have been published. The first considers a single-phase case with liquid-phase transport only. The second considers a two-phase case with both liquid-phase and gas-phase transport.

The near-field release of the single-phase scenario (Fig. 3; Keesmann et al., 2005) is the latest update of the Early-Intrusion-Case of Storck et al. (1988). About 2,000 m³ of brine is assumed to intrude into the repository and dissolve the waste up to the solubility limits shown in Table 1. The brine is squeezed out of the repository due to the convergence of the elastically deforming rock salt. The exit point of the contaminated brine is the contact between Quaternary sediments and the Main Anhydrite (Hauptanhydrit) of the Zechstein salt dome at the borehole location GoHy3020 (Fig. 2) on the -250-m level.

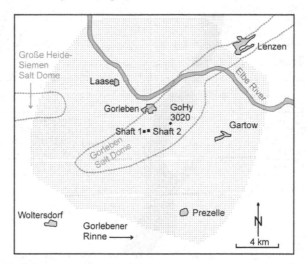

Fig. 2. Map of the Gorleben area with model mesh

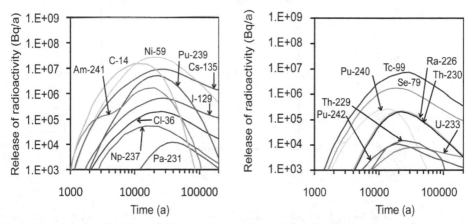

Fig. 3. Gorleben model. Near-field release of radionuclides (Bq/a) for the one-phase scenario

The two-phase scenario for the release of C-14 in the presence of a gas phase is based on a case described by Javeri (2006). Brine squeezed out from a 1,000 m³ brine pocket intrudes a backfilled disposal drift during 2,000 years following the closure of the repository. The disposal drift hosts 33 waste containers. At low oxygen fugacities, the container wall is subjected to corrosion dominated by the reaction

$$3Fe + 4H_2O = Fe_3O_4 + 4H_2. \tag{1}$$

The corrosion process releases hydrogen gas at a rate that increases from zero to 3.84 kg/a in the period from year zero to year 100 and decreases from 3.84 kg/a to zero in the period from year 20,000 to year 40,000. The hydrogen gas leaves the repository via the backfilled shaft according to the release rates shown in Figure 4.

Solubility (mol/L)					
	One-phase scenario	Two-phase scenario		One-phase scenario	Two-phase scenario
Am	1.0×10^{-5}	–	Pa	1.0×10^{-6}	–
C	1.0×10^{-2}	unlimited	Pu	1.0×10^{-6}	–
Cl	unlimited	–	Ra	1.0×10^{-6}	–
Cs	unlimited	–	Se	1.0×10^{-4}	–
I	unlimited	–	Tc	1.0×10^{-4}	–
Ni	1.0×10^{-4}	–	Th	1.0×10^{-6}	–
Np	1.0×10^{-5}	–	U	1.0×10^{-4}	–

Table 1. Gorleben model. Liquid-phase solubility in the near field

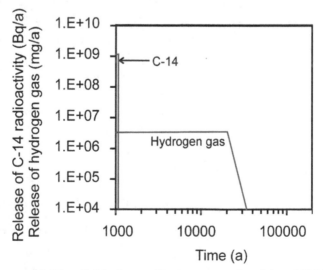

Fig. 4. Gorleben model. Near-field release of hydrogen gas (mg/a) and C-14 (Bq/a) for the two-phase scenario

Fig. 5. Gorleben model. Block diagram. Vertical exaggeration ratio 12:1

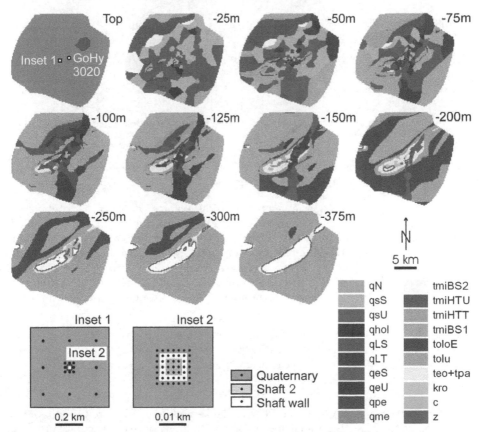

Fig. 6. Gorleben model. Hydrostratigraphic units (Table 3) of the top, -25 m, -50 m, -75 m, -100 m, -125 m, -150 m, -200 m, -250 m, -300 m and -375 m layer. The location of the borehole GoHy3020 and the structure of the mesh in the area surrounding Shaft 2 (Insets 1 and 2) are also shown

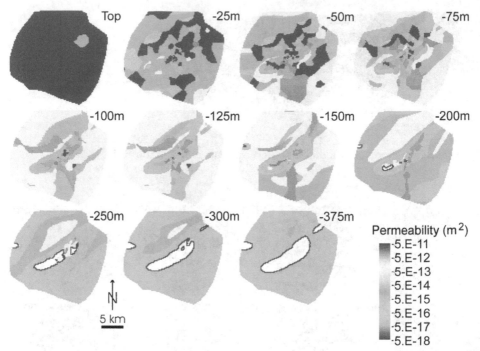

Fig. 7. Gorleben model. Permeability (m²) of the top, -25 m, -50 m, -75 m, -100 m, -125 m, -150 m, -200 m, -250 m, -300 m and -375 m layer

Together with hydrogen, C-14 species are released. The average waste canister contains C-14 with an activity of 3×10^{10} Bq (Keesmann et al., 2005). A fraction of 15 % is available for rapid release upon failure of the canister. This is assumed to result in a release of 13 % of the initial C-14 inventory out of the repository during a 100-year period starting in the year 1,000 after the closure of the repository. The corresponding release from the total of the 33 waste containers is 1.3×10^9 Bq/a. C-14 is the only radionuclide considered in the two-phase scenario because it is the only long-lived radionuclide that strongly partitions into the gas phase.

TOUGHREACT allows for injection of a single non-condensable gas, which is hydrogen in the case of the present study. Other gaseous species are introduced as secondary species via the corresponding primary liquid species according to a dissociation reaction such as

$$HCO_3^- + H^+ = CO_{2(gas)} + H_2O. \tag{2}$$

The primary species are HCO_3^-, H^+ and $O_{2(aq)}$. The secondary species are $CO_{2(gas)}$, $CH_{4(gas)}$, $CH_{4(aq)}$, $CO_{2(aq)}$ and CO_3^{-2}. The initial and boundary solutions have a H^+ activity of 10^{-7} and an $O_{2(aq)}$ activity of 10^{-67}. The source for the two-phase model is the near-field release shown in Figure 4. Hydrogen gas is injected into the primary central cell of shaft 2 on the -237.5-m level whereas a boundary solution with HCO_3^- is injected from a secondary boundary cell that is connected to this primary cell.

General properties	
Model length/width/height (m)	22,400/21,000/400
Model temperature (°C)	25
Simulation period (a)	200,000
Rock density (kg/dm³)	2.5

Saturated hydraulic properties (Table 3)	
Permeability (m²)	$5 \times 10^{-11} - 1 \times 10^{-18}$
Porosity (-)	0.03-0.3

Unsaturated hydraulic properties	
Residual liquid saturation	0.
Residual gas saturation	0.
Van Genuchten parameter α (Pa)	4.46×10^4
Van Genuchten parameter m (-)	0.21

Distribution coefficients (dm³/kg)					
	One-phase scenario				Two-phase scenario
	Low-salinity water (<10 g/L salt)		High-salinity water (≥10 g/L salt)		Low/high-salinity water
	Sand	Silt/clay	Sand	Silt/clay	Sand/silt/clay
Am	300	20,000	3,000	20,000	-
C	0.2	2	0.2	2	0.
Cl	0.1	0.1	0.1	0.1	-
Cs	70	400	2	70	-
I	2	2	0.1	2	-
Ni	20	300	6	90	-
Np	10	300	10	300	-
Pa	600	6,000	600	6,000	-
Pu	100	3,000	100	3,000	-
Ra	40	300	2	40	-
Se	1	1	1	1	-
Tc	1	6	1	1	-
Th	200	2,000	200	200	-
U	2	80	0.6	20	-

Aqueous phase diffusivity (m²/s)	
Salt	1.0×10^{-7}
Hydrogen	1.0×10^{-9}
All radionuclides	1.0×10^{-9}

Table 2. Set-up of the Gorleben far-field model

Quaternary			
Symbol	**Rock**	**Permeability (m^2)**	**Porosity (-)**
qN	Holocene, Weichsel glacial (sand)	5x10^{-11}	0.3
qsS	Saale glacial (sand)	4x10^{-11}	0.3
qsU	Saale glacial (silt, clay, drift marl)	1x10^{-14}	0.15
qhol	Holstein interglacial (silt, clay)	1x10^{-16}	0.03
qLS	Elster glacial/Lauenburger-Ton-Komplex (sand)	1x10^{-11}	0.3
qLT	Elster glacial/Lauenburger-Ton-Komplex (clay, silt)	1x10^{-16}	0.03
qeS	Elster glacial [excluding Lauenburger-Ton-Komplex] (sand)	1x10^{-11}	0.3
qeU	Elster glacial [excluding Lauenburger-Ton-Komplex] (silt, drift marl)	1x10^{-14}	0.15
qpe	Bavel-Cromer-Komplex (silt)	1x10^{-14}	0.15
qme	Menap glacial (sand)	1.5x10^{-11}	0.3
Tertiary/Cretaceous/Zechstein			
Symbol	**Rock**	**Permeability (m^2)**	**Porosity (-)**
tmiBS2	Lower Miocene/Obere Braunkohlensande (sand)	5x10^{-12}	0.3
tmiHTU	Lower Miocene/Hamburger-Ton-Komplex (silt, clay, sand)	1x10^{-14}	0.15
tmiHTT	Lower Miocene/Hamburger-Ton-Komplex (clay, silt)	1x10^{-16}	0.03
tmiBS1	Lower Miocene/Untere Braunkohlensande [includ. Neochat] (sand, silt)	5x10^{-13}	0.3
toloE	Upper Oligocene/Eochat (clay, silt)	1x10^{-14}	0.15
tolu	Lower Oligocene/Rupelton (clay, silt)	1x10^{-16}	0.03
teo+tpa	Eocene, Paleocene (clay, silt, sand)	3x10^{-14}	0.15
kro	Cretaceous (limestone, marlstone, sand)	1x10^{-14}	0.15
c	Caprock (former Zechstein salt)	5x10^{-13}	0.3
z	Zechstein salt	1x10^{-18}	0.1
Structural material			
	Permeability (m^2)	**Porosity (-)**	
Backfill of Shaft 2	3x10^{-13} - 3x10^{-12}	0.3	
Wall of Shaft 2	1x10^{-16}	0.1	

Table 3. Saturated hydraulic properties of rocks and structural material of the Gorleben far-field model

3.3 Geometry of the far-field model and initialisation

The irregular mesh has 290,435 elements and measures 22,400 m x 21,000 m x 400 m (Figs. 5-7; Table 2). Three sets of nodal distances in the x and y direction are used. The backfilled shaft 2, which measures 6 x 6 m horizontally, and the area immediately surrounding shaft 2 (shaft wall and bedrock) have horizontal nodal distances of 2 m. An intermediate zone (only bedrock) has horizontal nodal distances of 20 m. The remaining part of the primary mesh (only bedrock) has nodal distances of 200 m.

The top layer has nodal z values representing the head (between 12.7 m and 22.1 m altitude; Fig. 8). The second layer has nodes located 0.1 m vertically below those of the top layer. The vertical nodal distance between the following 32 layers is 12.5 m. The top layer, which serves to maintain constant pressure and salinity, exclusively consists of infinite-volume boundary elements (10^{45} m³). Other boundary cells are the rock-salt cells and the bottom-layer cells of the -375-m level. They have a volume of 10^{14} m³. This volume is large enough for maintaining nearly constant salinity throughout the simulation period but is simultaneously flexible enough to account for pressure adjustments. The volume of the remaining cells is calculated according to their nodal positions. A secondary boundary cell is connected to the primary cell in the centre of shaft 2 at the top of the salt dome (-237.5-m level). The boundary cell is used for injecting HCO_3^- whereas the primary cell is used for injecting hydrogen gas.

The initialisation starts with a pressure of 10^5 Pa in all models cells. The infinite-volume elements of the top layer have zero salinity. The large-volume elements (10^{14} m³) for the rock-salt unit and the bottom layer have an initial salinity of $X_{brine} = 1$ ($X_{salt} = 0.25$). Based on these boundary conditions, the diffusivity of brine is estimated in a trial-and-error procedure. Suitable vertical salinity gradients at the end of the 100,000-year initialisation period are obtained with a diffusivity 10^{-7} m²/s (Fig. 9). These salinity gradients, which are close to the present-day values (Klinge et al., 2007), are maintained with insignificant variations throughout the following 7,000-year and 200,000-year simulation periods.

3.4 Hydrostratigraphy and hydraulic properties

The hydrostratigraphic information of 197 drill hole logs is transferred to the nearest model node at the corresponding z-position (Figs. 6 and 8; Ludwig et al., 1989; Beushausen & Ludwig, 1990; Ludwig et al., 1993; Klinge et al., 2001). Based on this information, the hydrostratigraphic units for the remaining nodes are assigned by interpolation and extrapolation.

The saturated hydraulic properties of the hydrostratigraphic units of the caprock and cover sediments (Table 3) are assigned according to Vogel (2005). This scheme is based on the geometric mean of empirical maximum and minimum values. No decision has been taken yet with respect to the saturated hydraulic properties of the backfill of shaft 2. Therefore, various permeability cases are calculated (base-case permeability of 1 x 10^{-12} ; sensitivity cases of 3 x 10^{-13}, 5 x 10^{-13}, 2 x 10^{-12} and 3 x 10^{-12} m²). The porosity is 0.3 for all cases.

The unsaturated permeabilities are calculated with the relative-permeability function of Corey (1954; quoted by Brooks & Corey, 1964). The parameters for the capillary-pressure function of Van Genuchten (1980) are those of the Apache Leap tuff reported by Rasmussen (2001).

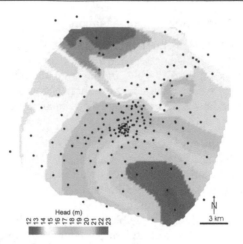

Fig. 8. Gorleben model. Head (m), location of Shaft 2 (open black square) and boreholes (solid black circles)

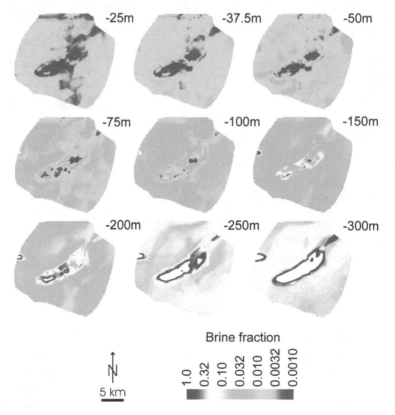

Fig. 9. Gorleben model. Salinity (brine fraction) of the -25 m, -37.5 m, -50 m, -75 m, -100 m, -150 m, -200 m, -250 m, and -300 m layer

	Inventory (Bq/t U)	Half-life (a)	Dose conversion factor (Sv/a:Bq/L)
Am-241 (parent of Np-237)	2.05×10^{14}	432	3.4×10^{-4}
C-14	1.89×10^{10}	5,700	5.9×10^{-5}
Cl-36	3.43×10^{8}	3.0×10^{5}	9.5×10^{-6}
Cs-135	2.18×10^{10}	2.0×10^{6}	3.7×10^{-6}
I-129	2.02×10^{9}	1.6×10^{7}	2.1×10^{-4}
Ni-59	6.02×10^{10}	7.5×10^{4}	5.4×10^{-8}
Np-237	1.34×10^{10}	2.1×10^{6}	7.5×10^{-4}
Pa-231	1.27×10^{6}	3.3×10^{4}	2.0×10^{-3}
Pu-239 (parent of U-235)	2.16×10^{13}	2.4×10^{4}	3.5×10^{-4}
Pu-240 (parent of U-236)	5.11×10^{13}	6.5×10^{3}	3.5×10^{-4}
Pu-242 (parent of U-238)	3.78×10^{11}	3.7×10^{5}	3.2×0^{-4}
Ra-226	1.06×10^{4}	1.6×10^{3}	1.8×10^{-4}
Se-79	1.75×10^{10}	1.1×10^{6}	6.1×10^{-7}
Tc-99	2.54×10^{11}	2.1×10^{5}	6.3×10^{-7}
Th-229	6.45×10^{3}	7.9×10^{3}	6.7×10^{-4}
Th-230 (parent of Ra-226)	3.92×10^{6}	7.5×10^{4}	1.1×10^{-4}
U-233	3.62×10^{6}	1.6×10^{5}	4.7×10^{-5}
U-235	4.97×10^{8}	7.0×10^{8}	4.5×10^{-5}
U-236	8.43×10^{9}	2.3×10^{7}	4.4×10^{-5}
U-238	1.22×10^{10}	4.5×10^{9}	4.2×10^{-5}

Table 4. Inventory of radionuclides in the candidate repository at Gorleben, half-life of radionuclides and dose conversion factor (DCF)

3.5 Sorption and diffusion

The distribution coefficients for the radionuclides in the one-phase scenario are taken from Suter et al. (1998). C-14 in the two-phase scenario is assumed to be non-sorbing under low oxygen fugacities compatible with the release of hydrogen gas. The aqueous phase diffusivity is 10^{-9} m^2/s for all species (Keesmann et al., 2005).

3.6 Simulated far-field transport

The simulation period is from year 1,000 to year 200,000 for the single-phase scenario and from year 1,000 to year 8,000 for the two-phase scenario. The input for the one-phase simulations is the near-field release of 17 radionuclides, of which five are the parents for daughter radionuclides (Am-241, Pu-239, Pu-240, Pu-242 and Th-230; Fig. 3; Table 4). The biosphere is represented by the -37.5-m level, which approximately is the limit between fresh water and saline water (1 g/L salt or 0.0034 brine fraction). The radioactive plume reaches its highest vertical extent in the year 200,000 (end of the simulation period) and in the year 7,000 for the one-phase and two-phase scenario, respectively (Figs. 10 and 11).

The calculation of the radioactive dose is based on Dose Conversion Factors (DCFs; Table 4; Hirsekorn et al., 1991). The annual dose D_D (Sv/a) is calculated according to the equation

$$D_D = \sum X_i C_i \ [\text{Sv/a}], \tag{3}$$

where X_i is the concentration (Bq/L) of the radionuclide i and C_i is the DCF (Sv/a:Bq/L) of the radionuclide i.

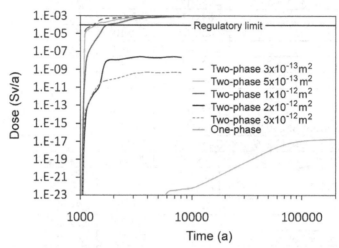

Fig. 10. Gorleben model. Peak values of the radioactive dose (Sv/a) on the -37.5 m level for the one-phase scenario and two-phase scenario (base case with a permeability of 1×10^{-12} m^2 for the backfill of Shaft 2 and sensitivity cases with a permeability of 3×10^{-13} m^2, 5×10^{-13} m^2, 2×10^{-12} m^2 and 3×10^{-12} m^2)

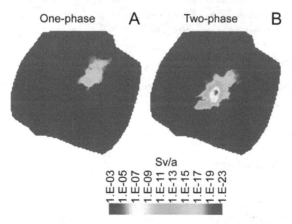

Fig. 11. Gorleben model. (A and B) Maps of the -37.5-m layer with radioactive dose (Sv/a) in the simulation year 200,000 for the one-phase scenario and year 7,000 for the base case of the two-phase scenario

For the single-phase scenario, the peak doses on the -37.5-m level is 1×10^{-17} Sv/a. For the two-phase scenario, the peak dose varies strongly depending on the assumptions for the permeability of the backfill of shaft 2. For the base-case permeability (1×10^{-12} m^2) and lower permeabilities, the peak dose is above the German regulatory limit of 1×10^{-4} Sv/a.

4. The granitoid-hosted repository at Olkiluoto

4.1 Repository layout

The planned repository is located in granitoid at 420 m depth at the island of Olkiluoto off the south-west coast of Finland (POSIVA, 2009). It can accommodate 3,000 containers with a total of 5,800 t U of burnt fuel (Löfman, 2005). According to the KBS-3V concept, the containers will be emplaced in vertical deposition holes driven from the floor of deposition tunnels, which are connected to a nearly horizontal system of access tunnels and transport tunnels. The containers that host the waste have a 5 cm thick corrosion-resistant copper shell around an inner deformation-resistant cast-iron canister; they are embedded in a 35 cm thick bentonite buffer. The horizontal dimensions of the repository are 1,600 m x 1,400 m.

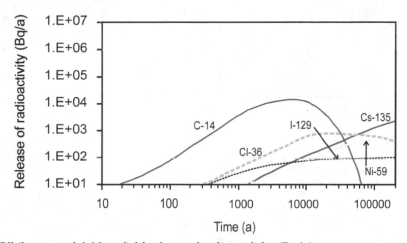

Fig. 12. Olkiluoto model. Near-field release of radionuclides (Bq/a)

4.2 Near-field

The near field is the hydrochemical regime of the underground workings and engineered barrier around the radioactive waste, consisting of the container wall and bentonite buffer. The reference failure mode for a defective canister and release of radioactivity from the near field to the undisturbed rock of the far field (geosphere) is the near-field base case (Nykyri et al., 2008). The canister is assumed to have a production defect with a diameter of 1 mm through which water enters into the canister causing dissolution of the fuel up to the solubility limits shown in Table 5. The near-field release of radionuclides (Fig. 12) is used as source for the contaminant injection cells in the far-field model.

As opposed to the easily corrodible steel canisters in the salt repository (see above), the copper shell of the canisters in the granitoid repository (KBS-3V concept) has an extraordinary chemical stability (Schwartz, 2008). Among the various candidate materials for waste packages, copper has unique oxidation characteristics. The conversion from native metal (Cu°) to metal oxide (Cu$_2$O) occurs in a mildly oxidising to mildly reducing environment (positive Eh for pH <7.7 at 25°C; Fig. 13) whereas the conversion of steel, titanium or nickel-chromium alloy already is possible under strongly reducing conditions, i.e., below the stability field of liquid water. Native copper is present as natural occurrence in the Earth's crust. The largest deposits are located on the Keweenaw Peninsula of Michigan, where they have been mined to a depth of 2.2 km (Schwartz, 1996). These deposits formed one billion years ago at temperatures of 100-200°C. The mineralising fluids were slightly less reducing but had much higher copper concentrations than common present-day groundwater.

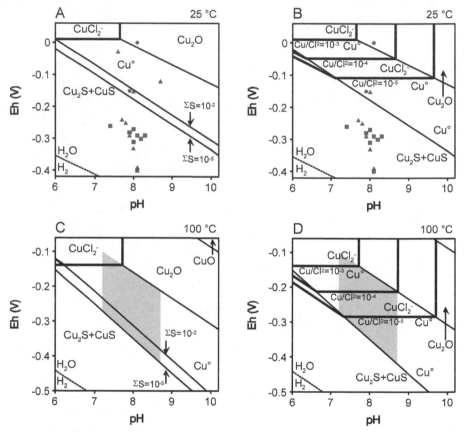

Fig. 13. Eh-pH diagrams for the system Cu-S-Cl-O-H at 25°C and 100°C for 0.1-0.3 mol/L Cl (Schwartz, 2008). Shaded area = Eh-pH range in which the Keweenaw native-copper deposits have formed (Schwartz, 1996). Point symbols = groundwater in crystalline rock of envisaged disposal sites in Sweden, Finland and Switzerland. (A and C) Diagrams for ΣS of 10^{-2} and 10^{-5}, and $\Sigma Cu/Cl^2 = 10^{-3}$. (B and D) Diagrams for $\Sigma S = 10^{-5}$ and $\Sigma Cu/Cl^2$ of 10^{-3}, 10^{-4} and 10^{-5}

Under the envisaged repository conditions, $CuCl_2^-$ is the dominant aqueous copper species. Accordingly, the thermodynamically relevant parameter is the $\Sigma Cu/Cl^2$ ratio, which refers to the sum of activities of all dissolved copper species divided by the Cl⁻ activity raised to the second power. Mine waters in the Keweenaw district have $\Sigma Cu/Cl^2$ ratios in the range from 10^{-4} to 10^{-5}. Similar conditions are achievable in the bentonite buffer around the copper container if the bentonite contains artificial admixtures of fine-grained native copper. This is a technically feasible variant of the KBS-3V concept, which simulates the environment of a natural native-copper deposit. Thus thermodynamic equilibrium between container and fluid on the high Eh side of the native-copper stability field is implemented in the repository-container system as analogue to the natural counterparts.

The low-Eh and low-pH side of the native-copper stability field is the copper-sulphide phase boundary (CuS and CuS_2). The position of this boundary depends on the total activity of aqueous sulphur species (ΣS). The groundwater in the envisaged repositories in Finland, Sweden and Canada has ΣS in the range from $<10^{-5}$ to 10^{-2}. This implies conditions partly below the stability field of native copper, pH being in the range of 7-9 and Eh in the range from -0.03 to -0.4 V. Provided the bentonite contains no sulphur-bearing minerals, the corrosion of the copper container is constrained by the diffusion of aqueous sulphide from the granitoid aquifer through the bentonite to the container surface. This situation is quite different from corrosion of other package materials such as steel, titanium or nickel-chromium alloy, which all corrode by reaction with omnipresent water. It is not possible to control corrosion of these materials by a diffusion barrier around the container.

The reference buffer of the KBS-3V concept is the pyrite-bearing MX-80 bentonite (0.3 % sulphide sulphur). Another option is the Avonlea bentonite, which has no sulphides but a high content of gypsum (2 %). Neither the presence of pyrite (FeS_2) nor gypsum ($CaSO_4 \cdot 2H_2O$) is desirable because ΣS of the system should be as low as possible. The Czech EKOBENT bentonite (Prikryl & Woller, 2002), which has a total sulphur concentration of less than 0.05 % and does not contain sulphides, would be the better choice although it has not been tested as thoroughly as MX-80 or Avonlea. The original idea behind the use of a sulphide-bearing bentonite in the KBS-3V repository was to accelerate the transition from oxidising conditions (during the construction phase) towards a final anoxic regime. This scheme, however, has been called in question (SKI, 2006). There are alternatives such as the implementation of saturated conditions in the shortest possible time. This can be achieved by incremental flooding of those parts of the slightly inclined emplacement tunnels where waste packages are already in place.

In the case of a sulphur-free buffer, general corrosion of the copper container can be determined by a simple calculation. The experimentally derived diffusion coefficient of aqueous sulphide in bentonite is approximately 7×10^{-8} cm^2/s (King & Stroes-Gascoyne, 1995; King et al., 2002). The groundwater of the envisaged repositories in Finland, Sweden and Canada has sulphide concentrations in the range from $<3 \times 10^{-7}$ to 3×10^{-5} mol/L. Assuming sulphide concentrations at the upper limit, 6×10^{-17} mol/(s cm^2) are transported through the 35 cm thick bentonite to the container surface. This is equivalent to a general corrosion rate of 3×10^{-8} cm/a, when Cu° is transformed to Cu_2S. Allowing for localised corrosion by multiplying the general rate with a correction factor of 5, the maximum corrosion depth is less than 0.2 cm within one million years. A service time for the 5-cm thick copper shell beyond one million years is a realistic possibility (Schwartz, 2008). The mass-transport assumption in this case sets an upper corrosion limit because some of the diffusing sulphide reacts with disseminated native-copper in the buffer and, therefore, is not available for corrosion of the copper container. Thus the near-field release for a canister that has no production defect and has not suffered mechanical damage is assumed to be zero for the 200,000-year simulation period.

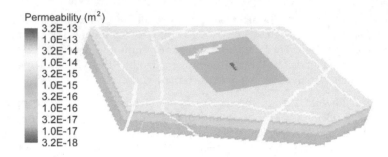

Permeability (m^2)
3.2E-13
1.0E-13
3.2E-14
1.0E-14
3.2E-15
1.0E-15
3.2E-16
1.0E-16
3.2E-17
1.0E-17
3.2E-18

Fig. 14. Olkiluoto model. Block diagram

C	unlimited
Cl	unlimited
Cs	unlimited
I	unlimited
Ni	4.3x10^{-3}

Table 5. Olkiluoto model. Liquid-phase solubility (mol/L) in the near-field

4.3 Geometry of the far-field model and initialisation

The model has a total of 323,657 elements. The irregular rectangular primary mesh is the fracture continuum, which measures 7,700 m x 5,500 m x 750 m (Figs. 14-16) and consists of 27 layers with 168,048 cubic or cuboid cells. The nodal distance in the x and y direction is 66.6 m. The top layer has nodal z values representing the head (-8 m to 10 m altitude). The second layer has nodes located 0.1 m vertically below those of the top layer. The vertical nodal distance between the following 25 layers is 30 m. The top layer, which serves to maintain constant temperature, pressure and salinity, exclusively consists of infinite-volume boundary elements (10^{45} m^3). The bottom layer cells have a volume of 10^{14} m^3. This volume is large enough for maintaining nearly constant temperature and salinity throughout the simulation period but is simultaneously flexible enough to account for pressure adjustments. The volume of the remaining cells is calculated according to their nodal positions. The secondary matrix continuum comprises the layers 2 to 26. The detailed geometric relationships between fracture and matrix continuum are determined on the basis of a parameter estimation procedure (see "Parameter estimation").

The infinite-volume elements of the top layer have a pressure of 10^5 Pa, a temperature of 6°C and zero salinity. The bottom-layer large-volume elements (10^{14} m^3) have an initial temperature of 14°C and an initial salinity of 0.025 mass fraction salt (corresponding to 0.44 mol/L Cl). The remaining layers have initial values along the gradients defined by the top and bottom layers. These boundary conditions and initial conditions, together with a high diffusivity of salt (10^{-3} m^2/s), allow to maintain suitable vertical gradients throughout the simulation, unless disturbed by radioactive heat sources. These gradients include temperature and salinity gradients that are close to the present-day values (Pitkänen et al., 2004; Löfman, 2005).

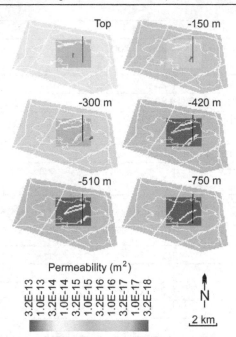

Fig. 15. Olkiluoto model. Permeability (m^2) of the top, -150 m, -300 m, -420 m, -510 m and -750 m layer. The position of the parameter-estimation model (black solid line orientated N-S) is also shown

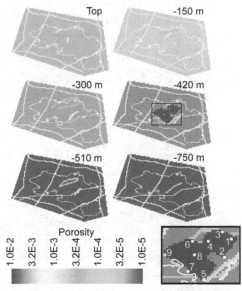

Fig. 16. Olkiluoto model. Flow porosity of the top, -150 m -300 m, -420 m, -510 m and -750 m layer. The repository area at the -420 m level is the red-coloured zone (indicating 0.01 porosity). Inset: The location of the source elements (1-9) in the repository area at the -420 m level

General properties	
Model length/width/height (m)	7,700/5,500/750
Simulation period (a)	200,000
Rock grain density (kg/dm³)	2.7
Rock grain specific heat (J/kg°C)	800
Formation heat conductivity (W/m°C)	3.65
Matrix porosity for anions (-)	0.001
Matrix porosity for neutral species and cations (-)	0.005
Flow porosity [average/range] (-)	$6.6 \times 10^{-5}/10^{-5} \text{-} 10^{-3}$
Fracture spacing (m)	5
Inventory	
C-14 (Bq/t U)	2.78×10^{10}
Cl-36 (Bq/t U)	1.04×10^{9}
Cs-135 (Bq/t U)	2.15×10^{10}
I-129 (Bq/t U)	1.14×10^{9}
Ni-59 (Bq/t U)	1.32×10^{11}

	Distance factor (-)		Retardation factor for the fracture continuum (-)	
	15-continua models	Dual-continuum models	15-continua models	Dual-continuum models
C-14	1	0.04	1	1
Cl-36	1	0.45	1	1
Cs-135	1	0.022	1	50
I-129	1	0.46	1	1
Ni-59	1	0.019	1	70

	Distribution coefficient (dm³/kg)	Effective diffusivity (m²/s)
C-14	0	1.0×10^{-13}
Cl-36	0	1.0×10^{-14}
Cs-135	50	1.0×10^{-13}
I-129	0	1.0×10^{-14}
Ni-59	100	1.0×10^{-13}

Table 6. Set-up of the Olkiluoto model

The simulation period is 200,000 years. During the first 10,000 years, the terrain is subjected to glacio-isostatic uplift (POSIVA, 2006; Löfman & Poteri, 2008) and the transient head conditions are translated into transient mesh geometries. The head reference point, which is the vertical projection of the north-western corner of the repository, is fixed at a constant value (2 m). The z values of the top layer change from simulation year zero to year 1,000 and 10,000 according to the z values in Figure 17. The rising terrain implies a retreating shoreline of the Baltic Sea, indicated by the <0-m head contours moving away from the present island of Olkiluoto.

There are nine contaminant-injection cells evenly distributed over the repository area on the -420 m level (Fig. 16). These source cells have only one connection to a fracture element. The near-field releases of the radionuclides shown in Figure 12 are used as input (see "Near field"). Transport paths and, hence, travel times for the radionuclides strongly depend on the location within the repository from which they are released. An ideal simulation would include all repository cells but this is too expensive computationally. Scoping calculations indicate that the relevant results such as median, mean and maximum doses change little when more than nine source cells are used.

4.4 Hydraulic properties

The geometry of the Hydrogeological Zones (HZs) of high permeability and surrounding Sparsely Fractured Rock (SFR) of low permeability corresponds to the hydrogeological structure model of Vaittinen et al. (2009). Permeability and flow porosity (Figs. 15 and 16) are based on POSIVA (2009). The only exception is the repository area on the -420 m level, where the permeability is based on Vaittinen et al. (2009) and POSIVA (2009) but the flow porosity is 0.01; this flow porosity value is derived from a backfilled repository volume of 1,310,000 m^3 (Kirkkomäki, 2007) and an average backfill porosity of 0.35.

4.5 Heat sources

The heat production of the radioactive waste in a single canister is shown for the first 2,000 years after deposition in Figure 18. These values are multiplied by a factor of 8.8, which is the number of canisters in a model cell, and are used as heat-generation data for the repository area on the -420 m level.

4.6 Parameter estimation

The retention of radionuclides in a fractured medium is caused by diffusion from the fracture into the stagnant fluid of the matrix adjacent to the fracture. An ideal simulation of a fracture media problem would require the discretisation of the matrix continuum (multiple-interacting continua or MINC model; Pruess, 1983; Pruess & Narasimhan, 1985; Pruess, 1992; Xu & Pruess, 2001; Liou, 2007).

Ten or more matrix continua are necessary for accurately simulating diffusive transport between fracture and matrix as the comparisons between numerical and analytical solutions have shown (Tang et al., 1981; Sudicky & Frind, 1982; Wu & Pruess, 2000; Diersch, 2009). However, the resulting problem size would be too large in the case of a three-dimensional simulation. Therefore, the problem size has to be reduced by replacing the model with a finely discretised matrix by an equivalent dual-continuum model (one matrix continuum). This is achieved by parameterising two variables among the three variables that control the retention of radionuclides; these are the diffusive transport distance between fracture and matrix, fluid volume of the fracture and effective diffusivity. The equivalent model is designed as conventional double-porosity model (Hantush et al., 2000, 2002; Lichtner, 2000).

The critical step in designing the model with a single matrix continuum (equivalent dual-continuum model) is the estimation of the parameters that justify to replace the corresponding finely discretised model (with a multi-continua matrix). The MINC pre-processor program of the TOUGHREACT code (Pruess et al., 1999) is suitable for models with up to 34 matrix continua. Scoping calculations have shown that models with 14 matrix continua have nearly the same retention properties as those of a model with the maximum number of continua. Thus, a two-dimensional 15-continua model has been designed that contains the fracture continuum of a 2,000-m long vertical subsection of the three-dimensional model enclosing the radioactive source element 1 (Figs. 15 and 16). A row of 14 matrix elements is connected to each fracture element with the exception of the fracture elements of the top and bottom layers.

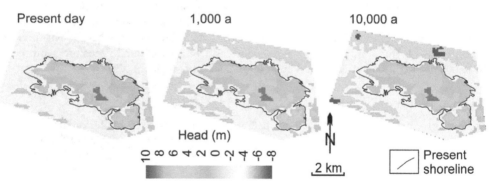

Fig. 17. Olkiluoto model. Vertical nodal positions (z coordinates; m) of the top layer in the simulation year zero, 1,000 and 10,000

The detailed geometry of the MINC mesh is calculated with the ONE-D proximity function of the MINC pre-processor program of the TOUGHREACT code. The fracture spacing is set at 5.0 m, which is the reciprocal of the weighted mean of the fracture intensity (Nykyri et al., 2008). The element volume fraction of the matrix continuum is derived from a matrix porosity of 0.001 for anions (Cl-36 and I-129) and 0.005 for neutral species and cations (C-14, Cs-135 and Ni-59) (Nykyri et al., 2008). According to the anion exclusion concept, anions and cations do not occupy the same pore system in low-porosity rocks, due to electrical forces caused by the usually negative charge of the mineral surfaces. Anions experience repulsion close to the mineral surfaces whereas cations are adsorbed on the surfaces (Rasilainen, 1997).

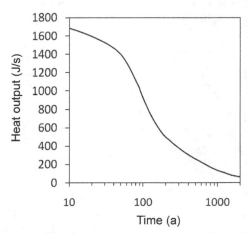

Fig. 18. Olkiluoto model. Heat output (J/s) from a canister with 1.93 t U burnt fuel from year 10 to year 2,000 after disposal (Löfman, 2005)

The near-field releases of radionuclides (Fig. 12) are used as input. The decay, diffusion and sorption data for the radionuclides (Tables 4 and 6; Nykyri et al., 2008) are the same as those for the three-dimensional far-field model.

Sorption of radionuclides takes place in the matrix continuum of the 15-continua model but not in the fracture continuum. The retardation factor $R_{m,i}$ of the radionuclide i (dimensionless) for the matrix continuum is calculated according to the equation

$$R_{m,i}=1+((1- n_m)K_{d,i}d)/n_m,$$ (4)

where n_m is the matrix porosity (dimensionless), d is the rock density (kg/dm³) and $K_{d,i}$ is the distribution coefficient (dm³/kg) of the radionuclide i for the rock matrix (Table 6).

TOUGHREACT calculates diffusion according to Fick's first law. The retention of radionuclides is controlled by the diffusivity, the fluid volume and transport distance between fracture and matrix. A dual-continuum model can almost perfectly substitute the 15-continua model if only two variables are parameterised; e.g., the fluid volume of the fracture together with the transport distance between fracture and matrix (as applied here) or the fluid volume of the fracture together with the diffusivity (producing similar results).

The fluid volume of the fracture continuum is adjusted by parameterising the retardation factor of the fracture continuum (which equals unity in the 15-continua model). The diffusive transport distance is parameterised by multiplying the fracture spacing of the 15-continua model (5 m) with a distance factor. The parameters are determined by the log-least-squares method. This method minimises the squared deviations between the annual far-field releases of the dual-continuum model and those of the 15-continua model (Fig. 19; Table 6). Note that the annual far-field release of a radionuclide (Bq/a) is the annual transport of the radionuclide from the second topmost layer to the topmost layer of the two-dimensional model.

The parameters obtained from this optimised two-dimensional model are used for the three-dimensional model that has one fracture continuum and one matrix continuum. This equivalent dual-continuum model is used for the subsequent transport simulations.

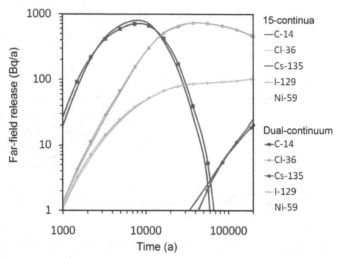

Fig. 19. Parameter estimation for the Olkiluoto model. The far-field release of radionuclides is shown for a 15-continua model and the equivalent dual-continuum model (optimised model)

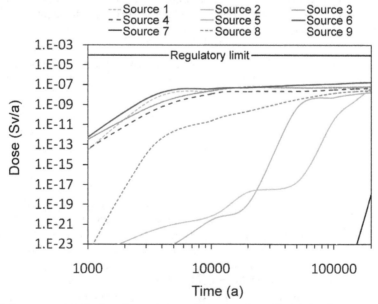

Fig. 20. Olkiluoto model. Peak values of the radioactive dose (Sv/a) for the second topmost layer for the sources 1-9

4.7 Simulated far-field transport

The radionuclide releases of the near-field base case are used as input for the far-field simulations. The biosphere is represented by the second topmost layer of the far-field model, which has fresh-water quality for most of the simulation period. The near-field base case assumes that one canister randomly chosen in the repository layout has a defect. There is a strong variability of radioactivity released to the biosphere depending on the location of the defective canister (Fig. 20). The canister locations 1-4, 6 and 8 in the north-eastern and central part of the repository imply a relatively large amount of radioactivity released to the biosphere whereas canisters located at the positions 5, 7 and 9 in the south-western part imply low or insignificant amounts of radioactivity. Figures 21A and 21B show the contaminant plumes for the source 3 at the NE corner and source 5 at the SE corner of the repository, respectively. The plumes emanating from the remainder of the canister locations have shapes that are intermediate between those shown. The maximum dose, which is produced by the canister position 6 (1.2×10^{-7} Sv/a), is far below the Finish regulatory limit of 1×10^{-4} Sv/a.

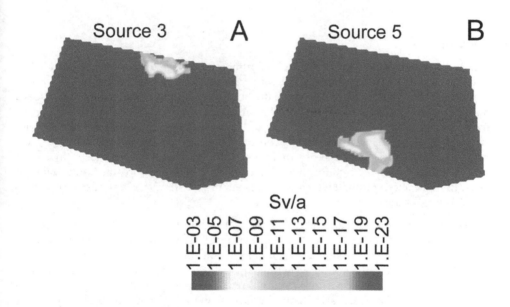

Fig. 21. Olkiluoto model. (A and B) Maps of the second topmost layer with radioactive dose (Sv/a) in the simulation year 200,000 for the sources 3 and 5

5. The Opalinus-Clay repository

The current scenarios assume that the repository will be located in the Jurassic Opalinus Clay in the Zürcher Weinland region, Switzerland (Fig. 1; NAGRA 2002). The claystone formation is gently dipping and has a thickness of 105-125 m. It is overlain by Middle Jurassic through Pleistocene sediments of approximately 600 m thickness. About 2,800 waste canisters containing 3,200 t U burnt fuel will be deposited in near-horizontal emplacement tunnels, which are connected to an access ramp via operational tunnels.

Two types of waste containers are considered for the repository. The first option is a steel canister, which is similar to that envisaged for the Gorleben repository. The second option is a container with a corrosion-resistant copper shell, such as the canisters for the Olkiluoto repository. So far, only the near-field release for the first option has been published (NAGRA, 2002; Kosakowski, 2004). For the second option, which considers improved safety standards, there are no near-field data available. The hydrochemical regime of a clay repository is similar to that of a granitoid repository but totally different from the highly corrosive regime of a salt repository. The performance of a copper-cast iron canister in a clay repository is as good as the performance of such a canister in a granitoid repository. Except for the case of a production defect or mechanical damage, the canister can assumed to be isolating the waste for a period of more than one million years. The permeability of a clay host rock is less than the permeability of a granitoid host rock. Thus it can be assumed that the potential groundwater contamination caused by a defective canister in the Opalinus-Clay repository is not higher than that for the Olkiluoto granitoid repository if the same type of waste canister is used.

6. Conclusions

The quality of models for groundwater contamination above high-level repositories in granitoid, salt and clay is not uniform. The most problematic case is a salt repository for two main reasons. First, the plastic behaviour of the host rock (salt) is a variable that introduces a high degree of uncertainty in any model attempt. Second, there is the inevitable generation of hydrogen gas caused by the corrosion of the steel waste canisters in high-salinity environments at low oxygen fugacities. The conditions of gas release constitute a great source of uncertainty. There is the possibility of unsteady or explosive release of hydrogen gas transporting short-lived radionuclides (Schulze, 2002). Furthermore, the permeability of the backfill of the shaft, through which the gas is assumed to flow, has a strong influence on the performance of the repository.

For a steady (non-explosive) gas release, a backfill permeability of $\leq 10^{-12}$ m^2 results in a radioactive dose of about 1 mSv/a. A higher permeability would decrease the radioactive dose to values below the German regulatory limit (0.1 mSv/a). A relatively high permeability favours the flow of liquid with respect to the flow of gas whereas a relatively low permeability has the opposite effect. A low permeability is the appropriate safety concept for a one-phase scenario (only liquid); however, a high permeability is the appropriate concept for a two-phase scenario (liquid and gas). As both scenarios are equally realistic, there is a typical conflict of targets. Such problems are not an important issue for the standard concept of a granitoid-hosted repository and the non-standard concept of a clay-hosted repository (both with copper-cast iron waste containers).

7. References

Beushausen, M. & Ludwig, R. (1990). Hydrogeologische Gliederung der oberoligozänen und miozänen Schichten. Bundesanstalt für Geowissenschaften und Rohstoffe, Hannover, Germany, Report Archive No. 106036, various paginations.

Brooks, R.H. & Corey, A.T. (1964). Hydraulic properties of porous media. *Hydrology Papers Colorado State University Fort Collins*, Vol. 3, pp. 1-27.

Corey, A.T. (November 1954). The interrelation between gas and oil relative permeabilities. *Producer's Monthly*, Vol. 19, No. 1 (quoted by Brooks and Corey 1964).

Diersch, H.J. (2009). *FEFLOW reference manual*. DHI-WASY GmbH, Berlin, Germany. URL: www.feflow.info/manuals.html

Hantush, M.M., Mariño, M.A. & Islam, M.R. (2000). Models for leaching of pesticides in soils and groundwater. *Journal of Hydrology*, Vol. 227, pp. 66-83.

Hantush, M.M., Govindaraju, R.S., Mariño, M.A. & Zhang, Z. (2002). Screening model for volatile pollutants in dual porosity soils. *Journal of Hydrology*, Vol. 260, pp. 58-74.

Hirsekorn, R.P., Nies, A., Rausch, H. & Storck, R. (1991). Performance assessment of confinements for medium-level and a-contaminated waste (PACOMA Project). GSF report 12/9 (GSF-Forschungszentrum für Umwelt und Gesundheit, Braunschweig, Germany), various paginations.

Javeri, V. (2006). Three-dimensional analyses of coupled gas, heat and nuclide transport in a repository including rock salt convergence. *Proceedings of the TOUGH Symposium 2006, 15-17 May 2006*, pp. 1-9, Lawrence Berkeley National Laboratory, Berkeley, California, U.S.A.. URL: http://esd.lbl

Keesmann, S., Nosek, U., Buhrmann, D., Fein, E. & Schneider, A. (2005). Modellrechnungen zur Langzeitsicherheit von Endlagern in Salz- und Granit-Formationen. GRS - Gesellschaft für Anlagen- und Reaktorsicherheit, Braunschweig, Germany, Report GRS-206, various paginations.

King, F. & Stroes-Gascoyne, S. (1995). Microbially influenced corrosion of nuclear fuel waste disposal containers. *Proceedings. Int. Conf. on Microbially Influenced Corrosion*, NACE International and American Welding Society, pp. 35/1-35/14.

King, F., Ahonen, L., Taxen, C., Vuorinen, U. & Werme, L. (2002). Copper corrosion under expected conditions in a deep geologic repository. *POSIVA Report* 2002-01. ULR: www.posiva.fi

Kirkkomäki, T. (2007). Design and stepwise implementation of the final repository (in Finnish, English abstract). *POSIVA Työraportty* 2006-92. URL: www.posiva.fi

Klinge, H., Boehme, J., Grissemann, C., Houben, G., Ludwig, R., Rübel, A., Schelkes, K., Schildknecht, F. & Suckow, A. (2007). Standortbeschreibung Gorleben Teil 1: Die Hydrogeologie des Deckgebirges des Salzstocks Gorleben. *Geologisches Jahrbuch*, Vol. C71, pp. 5-147.

Klinge, H., Margane, A., Mrugalla, S., Schelkes, K. & Söfner, B. (2001). Hydrogeologie des Untersuchungsgebietes Dömitz-Lenzen. Bundesanstalt für Geowissenschaften und Rohstoffe, Hannover, Germany, Report Archive No. 0121207, various paginations.

Kosakowski, G. (2004). Time-dependent flow and transport calculations for project Opalinus Clay (Entsorgungsnachweis). *NAGRA Technical Report* 03-10. URL: www.nagra.ch

Lichtner, P.C. (2000). Critique of dual continuum formulations of multicomponent reactive transport in fractured porous media. *Los Alamos National Laboratory Report* LA-UR-00-1097. URL: www.osti.gov

Liou, T.S. (2007). Numerical analysis of a short-term tracer experiment in fractured sandstone. *Terrestrial Atmospheric Oceanic Sciences*, Vol. 18, pp. 1029-1050.

Löfman, J. (2005). Simulation of hydraulic disturbances caused by the decay heat of the repository in Olkiluoto. *POSIVA Report* 2005-07. URL: www.posiva.fi

Löfman, J. & Poteri, A. (2008). Groundwater flow and transport simulations in support of RNT-2008 analysis. *POSIVA Working Report* 2008-52. URL: www.posiva.fi

Ludwig, R., Mandl, J. & Uhlig, A. (1989). Zwischenbericht über die Diskretisierung hydrogeologischer Strukturen im Bereich Gorleben. Bundesanstalt für Geowissenschaften und Rohstoffe, Hannover, Germany, Report Archive No. 0105061, various paginations.

Ludwig, R., Schneider, M. & Sewing, D. (1993). Projekt Gorleben: Hydrogeologische Gliederung der quartärzeitlichen Schichtenfolge. Bundesanstalt für Geowissenschaften und Rohstoffe, Hannover, Germany, Report Archive No. 110256, various paginations.

NAGRA (2002). Project Opalinus Clay - safety report. *NAGRA Technical Report* 02-05. URL: www.nagra.ch

Nykyri, M., Nordman, H., Marcos, N., Löfman, J., Poteri, A. & Hautojärvi, A. (2008). Radionuclide release and transport - RNT-2008. *POSIVA Report* 2008-06. URL: www.posiva.fi

Pitkänen, P., Partamies, S. & Luukkonen, A. (2004). Hydrochemical interpretation of baseline groundwater conditions at the Olkiluoto site. *POSIVA Report* 2003-07. URL: www.posiva.fi

POSIVA (2006). Expected evolution of a spent nuclear fuel repository at Olkiluoto. *POSIVA Report* 2006-05. URL: www.posiva.fi

POSIVA (2009). Olkiluoto site description 2008. *POSIVA Report* 2009-1. URL: www.posiva.fi

Prikryl, R. & Woller, F. (April 2002). Going underground – a new market for Czech bentonite in nuclear waste disposal. *Industrial Minerals*, pp. 72-77.

Pruess, K. (1983). GMINC - a mesh generator for flow simulations in fractured reservoirs. *Lawrence Berkeley National Laboratory Report* LBL-15227. URL: http://esd.lbl

Pruess, K. & Narasimhan, T.N. (February 1985). A practical method for modeling fluid and heat flow in fractured porous media. *Society Petroleum Engineers Journal*, pp. 14-26.

Pruess, K. (1992). Brief guide to the MINC-method for modeling flow and transport in fractured media. *Lawrence Berkeley National Laboratory Report* LBL-32195. URL: http://esd.lbl

Pruess, K., Oldenburg, C. & Moridis, G. (1999). TOUGH2 user's guide, version 2.0. *Earth Sciences Division, Lawrence Berkeley National Laboratory, University of California*, Berkeley, U.S.A.. URL: http://esd.lbl

Rasilainen, K. (1997). Matrix diffusion model. *VTT Publication* 331 (ESPOO, Helsinki, Finland). www.vtt.fi/inf/pdf/publications/1997/P331.pdf

Rasmussen, T.C. (2001). Pressure wave vs. tracer velocities through unsaturated fractured rock. *American Geophysical Union Geophysical Monograph*, Vol. 42, pp. 45-52.

Schulze, O. (2002). Auswirkung der Gasentwicklung auf die Integrität geringdurchlässiger Barrieregesteine. Bundesanstalt für Geowissenschaften und Rohstoffe, Hannover, Germany, Report Archive No. 0122442, 142 pp.

Schwartz, M.O. (1996). Native copper deposits and the disposal of high-level waste. *International Geology Review*, Vol. 38, pp. 33-41.

Schwartz, M.O. (2008). High-level waste disposal, ethics and thermodynamics. *Environmental Geology*, Vol. 54, pp. 1485-1488.

SKB (2006). Long-term safety for KBS-3 repositories at Forsmark and Laxemar - a first evaluation. *SKB Technical Report* TR-06-09. URL: www.skb.se

SKI (2006). Engineered barrier system - assessment of the corrosion properties of copper canisters. *SKI Report* 2006:11. URL: www/ski.se

Storck, R., Aschenbach, J., Hirsekorn, R.P., Nies, A. & Stelte, N. (1988). PAGIS performance assessment of geological isolation systems for radioactive waste: disposal in salt formations. *Office for Official Publications of the European Communities*, Luxembourg.

Sudicky, E.A. & Frind, E.O. (1982). Contaminant transport in fractured porous media: Analytical solutions for a system of parallel fractures. *Water Resources Research*, Vol. 18, pp. 1634-1642.

Suter, D., Biehler, D., Blasser, P. & Hollmann, A. (1998). Derivation of a sorption data set for the Gorleben overburden. *Proceedings DisTec'98, September 9-11, 1998*, pp. 581-584, Hamburg, Germany.

Tang, D.H., Frind, E.O. & Sudicky, E.A. (1981). Contaminant transport in fractured porous media: Analytical solution for a single fracture. *Water Resources Research*, Vol. 17, pp. 555-564.

Vaittinen, T., Ahokas, H. & Nummela, J. (2009) Hydrogeological structure model of the Olkiluoto site - update in 2008. *POSIVA Working Report* 2009-15. URL: www.posiva.fi

Van Genuchten, M.T. (1980). A closed-form equation for predicting the hydraulic conductivity of unsaturated soils. *Soil Science Society America Journal*, Vol. 44, pp. 892-898.

Vogel, P. (2005). Orientierende 3-D-Grundwassermodellrechnungen auf den Strukturen eines hydrogeologischen Modells Gorleben. Bundesanstalt für Geowissenschaften und Rohstoffe, Hannover, Germany, Report Archive No. 0125865, 70 pp.

Wu, Y.S. & Pruess, K. (2000). Numerical simulation of non-isothermal multiphase tracer transport in heterogeneous fractured porous media. *Advances in Water Resources*, Vol. 23, pp. 699-723.

Xiao, Z., Gammons, C.H. & Williams-Jones, A.E. (1998). Experimental study of copper(I) chloride complexing in hydrothermal solutions at 40 to 300°C and saturated vapor pressure. *Geochimica et Cosmochimica Acta*, Vol. 62, pp. 2949-2964.

Xu, T. & Pruess, K. (2001). Modeling multiphase non-isothermal fluid flow and reactive geochemical transport in variably saturated fractured rocks: 1. Methodology. *American Journal Science*, Vol. 301, pp. 16-33.

Xu, T., Sonnenthal, E., Spycher, N. & Pruess, K. (2005). TOUGHREACT user's guide: a simulation program for non-isothermal multiphase reactive geochemical transport in variably saturated geologic media. *Earth Sciences Division, Lawrence Berkeley National Laboratory, University of California*, Berkeley, U.S.A.. URL: http://esd.lbl

Zhang, K., Wu, Y. & Pruess, K. (2008). User's guide for TOUGH2-MP - a massively parallel version of the TOUGH2 code. *Earth Sciences Division, Lawrence Berkeley National Laboratory, University of California*, Berkeley, U.S.A.. URL: http://esd.lbl

6

Particulate Phases Possibly Conveyed from Nuclear Waste Repositories by Groundwater

Constantin Marin
"Emil Racoviță" Institute of Speleology of the Romanian Academy,
Bucharest
Romania

1. Introduction

Among the most severe sources of potential concern in terms of natural environment contamination with antropogenic radionuclides, a huge potential is associated to the radioactive waste resulting from the processing of the nuclear fuel used for energy generation purposes and from military and medical actvities. Although it represents about 3% of the total radioactive waste existing nowadays in the world, its total radioactivity amounts to more than 95% of that of the low & intermediate level and high-level radioactive waste taken together (Hu et al., 2010). By the present time, deep geologic disposal is considered to be the safest method for isolating highly radioactive and long-lived waste over millennia (Acero et al., 2010; Kim et al., 2011; Kurosawa & Ueta, 2001), while for the low and intermediate radioactive waste management, near surface storage is, for economic reasons, the preferred option (Dogaru et al., 2010; IAEA, 2004; Niculae et al., 2009).

It is a well-known fact that groundwater represents the most effective agent by which radionuclides stored in repositories could be transferred to the adjacent environment (Altmann, 2008; Baik et al., 2009; Geckeis & Rabung, 2008; Wersin et al., 2011). This is the reason why, each time when the issue of the radioactive waste repositories is addressed, irrespective of the repository type or of its inventory of stored radionuclides, the primary concern is to assess the local groundwater contamination risk.

Radionuclides may be conveyed by groundwater in a dissolved state, but especially in association with particles of various origins that are carried along with water. Those particles may have various structures and chemical compositions, their possible natures being: inorganic, organic or even micro-organisms. The inorganic particles genesis is closely related to the geochemistry of the environment with which groundwater is in contact, while particles of organic and biological nature are, as a general rule, allochtonous.

Particles dimensions may be very different, ranging from those of the colloids, up to microscopic particles, generically designated as "suspended particulate matter". Their reactivity and their ability of binding and conveying contaminants is to a certain extent controlled by their dimension, namely their reactivity is progressively enhanced as the particulate phase is more finely dispersed (Kersting & Zavarin, 2011; Wigginton et al., 2007).

This is also the reason why an increased interest is dedicated to the radionuclides behaviour in the presence of the colloidal matter.

The role played by the colloidal phase is intensely investigated not only for groundwater, but also for terrestrial surface water, namely for streams and lakes (Aleksandrova et al., 2010; Bondareva, 2011; Matsunaga et al., 2004; Monte, 2010; Monte et al., 2009; Ollivier et al., 2011; Semizhon et al., 2010), for estuarine water (Barros et al., 2004; Eyrolle & Charmasson, 2004; Porcelli et al., 1997), for sea water (Bowie et al., 2010; Otosaka et al., 2006; Scholten et al., 2005) or for soil water (Goryachenkova et al., 2009; Maity et al., 2011; Matisoff et al., 2011; Rachkova et al., 2010; Seliman et al., 2010; Xu et al., 2011). The colloid-radionuclide interaction is important not only from the perspective of their migration. For instance, the colloidal systems build-up is utilized for the radionuclides separation (Mansur & Mushtaq, 2011) and there has been noticed that they have interfering effects in the quantitative determination of the radio-isotopes. (Constantinou & Pashalidis, 2010; 2011; Kiliari & Pashalidis, 2010; Kyriakou & Pashalidis, 2011).

The radionuclides migration associated with the particulate phase in general, and with colloids in particular, that flow along with groundwater, is a topic of utmost importance for assessing the safety of the radioactive waste repositories. For assessing that overall process, there are conducted both field observations and experimental simulations in laboratory. The *in situ* investigation and the modeling of the role played by the colloidal matter and by the suspended particulate matter are mainly concerned with the saturated region of the water-bearing structures (e.g., Baik et al., 2010; Grambow, 2008; Kelkar et al., 2010; Laverov et al., 2010; Mal'kovskii, 2011; Malkovsky, 2011; Mazurek et al., 2011; Pourret et al., 2010; Severino et al., 2007; Utsunomiya et al., 2009). The modeling of the radionuclides migration in the unsaturated zone is addressed, as a general rule, by laboratory experiments (e.g., Ku et al., 2009; Massoudieh & Ginn, 2007). The latter range into two broad categories, batch tests and flow-through column experiments. In the first case, a solution spiked with the investigated radionuclide is mixed for a certain time-interval with the solid of interest, then the solution and/or the solid are analyzed (e.g., Anderson et al., 2009; Bradbury et al., 2005; Hu, Cheng, Zhang & Yang, 2010; Lujanienė et al., 2010; Rabung et al., 2005; Singer, Maher & Brown Jr, 2009). The column experiments investigate the radiocolloids migration characteristics, by simulating the groundwater flow conditions (e.g., Bryan et al., 2005; Li et al., 2011; Mibus et al., 2007; Solovitch-Vella et al., 2006).

Radionuclide migration with groundwater colloids through porous media (e.g., Bradford & Bettahar, 2006; Delos et al., 2008; Grolimund et al., 2001; Grolimund & Borkovec, 2001; 2006; Ilina et al., 2008; Kretzschmar et al., 1997; Li et al., 2010; Panfilov et al., 2008; Santos & Barros, 2010), or across fractured rock systems (e.g., Hu & Mori, 2008; Jeong et al., 2011; Kosakowski, 2004; Kurosawa & Ueta, 2001; Malkovsky & Pek, 2009b; Schindler et al., 2010; Tang & Weisbrod, 2009; 2010; Yamaguchi et al., 2008) are intensely investigated topics. At the same time, extensive research is conducted for assessing the humic and fulvic colloids effect on the radionuclides migration across the underground environment (e.g., Bouby et al., 2011; Geraedts & Maes, 2008; Joseph et al., 2011; Lippold et al., 2005; Lippold et al., 2005; Pshinko, 2009; Pshinko et al., 2009; Schmeide & Bernhard, 2010; Singh et al., 2009; Singhal et al., 2009; Yoshida & Suzuki, 2006).

The present work intends to succinctly review and to critically analyze the most recent contributions of the scientific literature in which there are addressed certain issues concerning the part that particulate phases play in the migration of the radionuclides characteristic to the radioactive waste stored in geological repositories, across more or less deep aquifer structures. An attempt is performed to provide an adequate definition of the particulate phases, and the main issues related to the radionuclides speciation are discussed, with particular emphasis on the radiocolloids development and distribution. A topic that is widely addressed in the present work concerns the radioparticles fractionation and their chemical characterization.

2. Radioparticles in groundwater

2.1 Particulate phases

It is a well-known fact that in a natural subsurface water-rock system, elements are distributed among the three constitutive phases, namely: (1) the solid phase, consisting of the mineral substratum and of the sediments; (2) the aqueous phase, within which elements are considered to occur as dissolved species, and (3) the particulate phase. It is by now fully demonstrated and unanimously accepted that much higher concentrations of contaminants are being carried by the particulate phases, as compared to concentrations of the corresponding dissolved species which are carried in the aqueous solution (Kalmykov & Denecke, 2011; Morel, 1983; Stumm & Morgan, 1996).

The particulate phase conveyed by groundwater is highly dynamic. Particles are continuously produced as a result of physical erosion and of chemical alteration of the mineral substratum, or ensuing to precipitation from super-saturated solutions. They undergo permanent composition changes, and are continuously removed from the water by dissolution, coagulation, and finally by deposition or binding to the solid phase.

Particles existing in groundwater may be both of inorganic and of organic nature (Wolthoorn et al., 2004). As a general rule, the inorganic particles composition mirrors to a large extent the nature of the mineral substratum in contact with water. Specifically, those particles may include solid fragments dislocated from the substratum, clay minerals, Fe(III), Mn(III,IV) and Al(III) oxihydroxide microparticles, silicates, carbonates, complexes and polymers of certain elements, etc. Organic particles are, at least in principle, of allochtonous origin, and they include fragments of degraded organic matter, macromolecules of organic substances which are specific to the soil (ex. humic or fulvic acids), but also organisms which are either alive or decomposing, microorganisms, viruses, as well as their exudates.

Aquatic particulate phases exhibit a continuous particle size distribution. In spite of that, practical reasons require that distinction is made between constituents that are dissolved, and those existing as particles on the one hand, and the various types of particles on the other. From a thermodynamic perspective, "dissolved" refers to a constituent for which a chemical potential can be defined (Stumm & Morgan, 1996).

In terms of dimension, shape and characteristics of transfer across the environment, taken as a whole, the particles range in two broad categories, namely colloids and suspended particles. A distinction between those two categories made by taking into account only the dimension criterion fails to be entirely satisfactory, although it is widely accepted and operationally useful. Opinions fail to be unanimous even as far as the dimensional boundary separating the two categories is concerned, so that according to various authors, the corresponding dimension may be $0.20\,\mu m$, $0.45\,\mu m$ or $1\,\mu m$ (Malkovsky & Pek, 2009a). In fact, the indicated values represent the pore dimensions of the filtering membranes which are used for separating the particles. The lower boundary of the colloidal particles dimension is accepted to be, as a general rule, $\approx 1\,nm$, a circumstance which authorizes, from a certain perspective, the assimilation of colloids to nanoparticles (Geckeis et al., 2011; Wigginton et al., 2007). An important characteristic of the aquatic colloids, useful in distinguishing them from the suspended particles, consists in the fact that their vertical movement is not significantly affected by gravitational settling (Gustafsson & Gschwend, 1997; Stumm & Morgan, 1996).

2.2 Speciation of radionuclides in the environment

The migration of an element in a natural environment is basically conditioned by the way it speciates in that environment. By definition, the chemical species of a certain element are its specific appearances, defined as electronic or oxidation states, as complex or molecular structures (Templeton et al., 2000). At a given moment, the same element may occur in the hydrosphere under various physical-chemical appearances, including forms associated to the particles in suspension or dissolved forms, such as simple inorganic species, organic complexes, metallic ions adsorbed to the colloidal matter, etc.

Accounting for the radionuclides speciation is a fundamental step toward describing the geo- and bio-chemical processes in which they are involved, and above all for understanding how they migrate through the environment. (Hu et al., 2010; Salbu, 2009; Salbu et al., 2004). The definitions of the terms addressing the elements speciation, recommended by International Union for Pure and Applied Chemistry (IUPAC) (Templeton et al., 2000), have been adapted for the radionuclides by Salbu & Skipperud (2009) as follows:

Radionuclide species are defined according to their physicochemical properties such as nominal molecular mass, charge properties and valence, oxidation state, structure and morphology, density, degree of complexation.

The *speciation of radionuclides* is the distribution of a radionuclide amongst defined chemical radionuclide species in a system.

Colloids and suspended particles cannot be considered, under the above definition, species, they are, alternatively, chemical fractions. The *chemical fraction* is a group of chemical entities which have common physical (e.g. size), or chemical (e.g. reactivity) properties, that group being operationally outlined by means of an analytical process. The concerned analytical process is termed *fractionation*. Chemical fractions are not mutually exclusive, i.e. rather than identifying the involved chemical species, they indicate a specific behaviour. In terms of involved experimental techniques, fractionation is much more accessible that the detailed determination of chemical species. For instance, the results of the *in situ* fractionation of water samples derived from The Fen Central Complex (southern Norway), one of the world's largest natural reservoirs of thorium (^{232}Th), have shown that radionuclides occur mainly as colloids and chemical species of low molecular mass (Popic et al., 2011).

There is a certain dependence of the chemical reactivity of particles carried by water on their dimension (Wigginton et al., 2007), namely reactivity is, as a general rule, enhanced, as the particle dimension decreases. This behaviour is mainly a result of the fact that with decreasing particle dimension, the ratio between the constitutive atoms and the surface area of the particle increases. This is one of the reasons why a large interest is being recently dedicated to the investigation of how radionuclides are transferred by means of colloidal particles. At the same time, the colloids mobilization and transfer can occur both in a saturated, and in an unsaturated flow regime, this quite important issue requiring a careful consideration for an appropriate management of the radioactive waste repositories safety. The part that suspended particles, and macro-particles in general, play in the radionuclides migration is not a negligible one, yet it is worth considering it only in a saturated flow regime.

3. Formation of the radiocolloids

Entities formed by the coupling of colloidal particles with radionuclides and conveyed as such by groundwater, and which commonly are also termed *radiocolloids*, are classified in two large groups, according to their origin: (1) *intrinsic-*, *eigen-colloids* or "true" colloids; and (2) *carrier-* or *pseudo-colloids* (Geckeis et al., 2011; Malkovsky et al., 2009; Malkovsky & Pek, 2009a).

3.1 Intrinsic radiocolloids

Intrinsic-colloids are formed spontaneously, as a result of the polymerization of complexes derived from the hydrolysis of metal ions. Under certain environmentally-controlled circumstances, an increased tendency to form such colloids is displayed by Pu, Am, Np, for which the dimensions of the resulting aggregates and their number per unit volume are, as a general rule, proportional to the total concentrations of the concerned actinides in the solution (Murakami et al., 2005).

It is a well-known fact that tetravalent plutonium has a strong predilection to develop polymeric complexes and colloids. There has been however noticed that small polymers such as dimers, trimers and tetramers include mixed oxidation states of Pu (Walther et al., 2009). In natural waters with pH ranging between 6 and 8, Pu(IV) prevalently occurs under the form of true colloids. The latter exhibit a predisposition for getting bound to the rocks surface, a behaviour which with increasing particles dimension, becomes more and more obvious. It was found, for instance, that when the particles dimensions exceed 220 nm, Pu(IV) is virtually quantitatively sorbed on the rock surface (Perevalov et al., 2009). At the same time, equilibrium distribution of Pu(IV) polymers depending on the total Pu(IV) concentration in the solution was analyzed theoretically by Kulyako et al. (2008).

3.2 Pseudo-colloids

Pseudo-colloids are formed through the binding of radionuclides to the pre-existing colloidal particles of the groundwater. Any mineral fragment, either crystalline or amorphous (hydrated Al, Fe and Mn oxides), organic compounds (humic and fulvic acids), but also biota consisting prevalently of viruses and bacteria may act as carrier particles for the radio-nuclides. As compared to the first group of radiocolloids, pseudo-colloids are much more abundant in groundwater, and therefore they exert a much more extensive control on the radionuclides transfer.

Experiments have been conducted which were aimed at establishing the conditions under which Fe(III), Cr(III) and Zr(IV) build up pseudo-colloids together with colloidal silica. There was thus noticed that in a Fe(III) solution with a concentration of $\approx 1 \times 10^{-7}$ M, under pH conditions normally met in the hydrosphere, Fe^{3+} cations and mononuclear $Fe(OH)_n^{3-n}$ hydroxo complexes mostly occur, and pseudo-colloids may form by the binding of Fe(III) species to colloidal silica (Davydov et al., 2003). In Cr(III) $\approx 1 \times 10^{-6}$ M synthetic solutions, when pH >4, in solution there are prevailing the $Cr(OH)^{2+}$ and $Cr(OH)_2^+$ species, with chromium displaying an obvious predilection toward forming pseudo-colloids with silica (Davydov et al., 2006). At the same time, in a solution of $\approx 1 \times 10^{-13}$ M concentration, Zr(IV) occurs under hydrated form as $Zr(OH)^{3+}$ and $Zr(OH)_2^{2+}$. At pH 2–12 Zr(IV) participates in formation of stable pseudocolloid particles (Davydov et al., 2006).

By using the surface complexation model, several investigators have modeled the radionuclides adsorption on the surface of the colloidal particles to form pseudo-colloids. For instance Batuk et al. (2011) interpreted in this way the sorption behavior and speciation of U on silica colloids, Degueldre & Bolek (2009) modeled plutonium adsorption on hydrous metal oxide solids, Del Nero et al. (2004) interpreted Np(V) sorption on amorphous Al and Fe silicates, and uranyl ions on Al-hydroxide (Froideval et al., 2006). Sorption of Np(V), Pu(V), and Pu(IV) on colloids of Fe(III) oxides and hydrous oxides and MnO_2 was studied over wide ranges of solution pH and ionic strength by Khasanova et al. (2007). The surface complexation model assumes that the adsorbing ion forms a surface complex with the adsorbing site, similar to the formation of a dissolved complex.

A box model has been proposed in order to interpret the kinetics of the radionuclides uptake on suspended particulate matter (Barros & Abril, 2005; 2008). At the same time, there has been investigated the kinetics of the Cs(I) sorption on hydrous silica (Pathak & Choppin, 2006), and Am^{3+} on suspended silica as a function of pH and ionic strength in the presence of complexing anions, humic acid and metal ions (Pathak & Choppin, 2007).

Living organisms with dimensions similar to colloids, like for instance pathogenic bacteria or viruses, are present in groundwater naturally, to form a distinct group named biocolloids (Bekhit et al., 2009). Since they are living organisms they migrate in the subsurface porous medium, being subject to a complex of biological, physical and chemical processes. Those micro-organisms act as pseudo-colloids, since their surfaces are often negatively charged, thus having the ability to bind and carry radionuclides through the subsurface environment (Johnsson et al., 2008; Luk'yanova et al., 2008; Seiler et al., 2011; Singer, Farges & Brown Jr, 2009; Wilkins et al., 2006; 2010).

3.3 Colloids generated by engineered barriers

Besides the two already mentioned colloid groups, Malkovsky et al. (2009) and Malkovsky & Pek (2009a) distinguish an additional third group, which they designate as "primary colloids". The latter are colloidal particles derived as a result of groundwater leaching the isolating materials utilized in the storage of low and intermediate level radioactive wastes, as well as of high-level radioactive wastes in geological disposal.

Action taken in order to prevent, as much as possible, the contamination of the geological environment with radionuclides stored in a radioactive waste repository, irrespective whether the latter is located at the ground surface or in the underground, makes use of the so-called *"barriers"*. The broadest meaning for barrier is "a physical obstruction that prevents or delays the movement of radionuclides or other material between components in a system, for example a waste repository" (IAEA, 2003). Such barriers may be either natural (i.e. geological), or constructed, in that latter case being called *"engineered barriers"*. In most instances there are used *"multiple barriers"*, namely "two or more natural or engineered barriers used to isolate radioactive waste in, and prevent migration of radionuclides from, a repository" (IAEA, 2007). Such barriers include the glass or ceramic matrixes to which the liquid waste is usually converted (Anderson et al., 2009; Curti et al., 2009), the steel canisters used for isolating the waste, the buffering backfill materials placed between the containers and the walls of the repository cells. The most adequate filling material for this purpose is bentonitic clay (Akgün et al., 2006; Ferrage et al., 2005; Galamboš et al., 2009; 2011; Gaucher et al., 2004; Pérez del Villar et al., 2005), since it has a low permeability and its coefficient of radionuclides diffusion is quite small (Arcos et al., 2008; Bradbury & Baeyens, 2011; Hu, Xie, He, Sheng, Chen, Li, Chen & Wang, 2010; Missana & García-Gutiérrez, 2007; Wang et al., 2005). When in contact with groundwater, all those barriers can release colloidal particles that in terms of both their chemical, and their mineralogical composition, are not characteristic to the concerned geological environment (Cadini et al., 2010; De Windt et al., 2004; Filby et al., 2008; Kunze et al., 2008; Wieland et al., 2004).

The effect of engineered barriers in terms of radiocolloids production is an outstandingly important research topic. It is worth mentioning that especially bentonite barriers in contact with weakly mineralized groundwater generate a highly concentrated colloidal phase which is liable to carry radionuclides (Albarran et al., 2008; 2011; Kalmykov et al., 2011; Kurosawa & Ueta, 2001; Missana et al., 2008; Sabodina et al., 2006; Tertre et al., 2005; Vilks et al., 2008).

3.4 The colloids stability

The colloids stability in groundwater is primarily controlled by the processes through which they agglomerate; at their turn, those processes are ruled by the colloid surface charge and by the solution composition, namely by its pH and ionic strength. (Geckeis et al., 2011; Schelero & von Klitzing, 2011). Accordingly, the groundwater chemistry plays a fundamental part in controlling the stability of the colloidal particles. Those particles ability of remaining in suspension in an aqueous environment depends on the interactions that are established between them when they reach close to one another. Colloids become stabilized through the formation of an electric double layer strong enough for preventing agglomeration. Yet this layer may be destroyed and the colloidal particles consequently coagulate and leave the system, along with the increase in ionic strength. There has been noticed that a reverse correlation exists between the colloids concentration in the solution and the ionic strength of the latter (Deepthi Rani & Sasidhar, 2011; Loux, 2011).

4. Fractionation and radioparticles charcterization

Analytical tools and detection methods used to characterize radioparticles in groundwater may be categorized as a function of the parameter to be determined as follows: (1) size fractionation; (2) size distribution; (3) surface area characterisation; (4) chemical and radiochemical analysis (May et al., 2008). Common size fractionation methods include ultrafiltration, tangential/cross-flow ultrafiltation (TFF/CFF), centrifugal-ultrafiltration (e.g., Gimbert et al., 2005; Liu et al., 2006; Pourret et al., 2007) and field-flow fractionation (FFF). Normally, those techniques are followed by the chemical analysis of the separated fractions. They will be discussed in detail in the following sections, since they are, taken together, the most frequently used experimental investigation approaches.

The size distribution of radioparticles can be investigated by a wide variety of techniques, such as laser light scattering (LLS) (e.g., Dreissig et al., 2011), diffuse light scattering (DLS) (e.g., Lahtinen et al., 2010), laser-induced breakdown detection (LIBD) (e.g., Baik et al., 2007), or photoelectron spectroscopy (Laverov et al., 2010). For the same purpose, atomic force microscopy (AFM) and transmission electron microscopy (TEM) (e.g., Doucet et al., 2005; 2004) are used as well.

The specific surface area (SSA) of the particles is the parameter describing the interdependence between the particles dimensions and their chemical or mineralogical composition. The SSA determination is frequently conducted by means of the Brunauer-Emmett-Teller (BET) gravimetric method. In an indirect way, information about the particles surface area may be obtained by means of AFM or TEM measurements.

In order to characterize the radionuclides speciation, a series of investigators have resorted to sequential chemical extraction experiments, conducted in accordance with pre-established work protocols (Bondareva, 2011; Bondareva & Bolsunovskii, 2008). As a general rule, ensuing to a scheme of sequential extraction of radionuclides from colloidal matter of groundwater, the following products will result: (1) water-soluble; (2) exchangeable, by using as reactant a 0.5 M $Ca(NO_3)_2$ solution at pH 5.5; (3) associated with carbonates, in the presence of 0.1 M NH_4Ac solution, pH 4.8; (4) associated with organic matter, with 0.1 M NaOH solution, pH 10; (5) amorphous oxides, by using a mixture of 0.18 M $(NH_4)_2C_2O_4$ and 0.1 M $H_2C_2O_4$ solutions, at pH 3.5; (6) the residue digestion by means of HF (Novikov et al., 2009; Novikov et al., 2009).

4.1 Suspended particulate matter (SPM) fractionation

The SPM analysis techniques have been developed and are frequently conducted for surface, estuarine or sea water samples, but they are equally utilized as well for the analysis of radionuclide-contaminated groundwater (Katasonova & Fedotov, 2009; Stepanets et al., 2009).

Two different approaches are utilized in order to determine the concentrations of heavy metals and radionuclides bound on the SPM, one which is direct, and the other indirect. The direct determination method consists in separating the suspensions on filtering membranes of various porosities, followed by subsequent digestion and quantitative assessment of the contaminants from the separated material, by means of an adequate spectrometric technique (Blo et al., 2000; Nordstrom et al., 1999; Ödman et al., 1999; 2006; Ollivier et al., 2011; Yeager et al., 2005). Through the indirect method, both the filtered and the unfiltered water samples are analyzed in parallel, and the resulting concentration difference is considered to represent the concentration of the element bound on the SPM (Cidu & Frau, 2009; Cortecci et al., 2009; Gammons et al., 2005; Pokrovsky & Schott, 2002).

In a comparative study, Butler et al. (2008) have demonstrated that congruent results were obtained when the two methods were applied in parallel. Potential artifacts induced by filtration, such as contamination and/or adsorption of metals within the membrane have been investigated for different membrane materials, metals, etc. (Hedberg et al., 2011).

4.2 Ultrafiltrafiltration

Ultrafiltration became the usual technique for separating colloidal particles from any type of natural water. The separation is performed by using a filtering membrane of a nominal size, often reported in a molecular size cut-off, in dalton (Da) units. Dalton is non-SI unit accepted for being utilized in the International System of Units whose values in SI units must be obtained experimentally. One dalton unit is equivalent to one atomic mass unit (amu) and is used in ultrafiltration in order to determine the approximate size of particles for which a rigorous molecular mass cannot be indicated.

Single filtration is frequently used for radioparticles separation (Caron & Smith, 2011; Novikov et al., 2009). Many investigators resort however to sequential (cascade) filtration, which involves the utilization of a series of cells with mixing, where each cell contains a membrane and a drainage system connected to measuring devices (Bauer & Blodau, 2009; Dreissig et al., 2011; Eyrolle & Charmasson, 2004; Graham et al., 2011; Pourret et al., 2010; Stepanets et al., 2009). In most cases, before starting the ultrafiltration operation, the SPM is separated by filtering the water samples on filtering membranes of 0.45 µm or 0.22 µm (e.g., Graham et al., 2008; Novikow et al., 2009; Singhal et al., 2009).

Tangential-flow ultrafiltration (TFF/CFF) is a common method for size fractionation in natural waters which has also been applied for colloids separation (e.g., Andersson et al., 2001; Buesseler et al., 2009; Goveia et al., 2010; Hassellöv et al., 2007; Ohtsuka et al., 2006). The main advantage of TFF is its use as a preparative fractionation method that allows for processing of large volumes of sample - even water samples reaching, each one, up to several hundreds of liters. It is also well known that the size distribution of colloids in natural waters can easily change due to aging, changes in pH, ionic strength or redox conditions (Hedberg et al., 2011). The processes that can potentially alter the size distributions of the colloids include coagulation, adsorption to surfaces, hydrolysis and precipitation. In addition to these processes, associated trace constituents are also affected by sorption processes, solution complexation and redox precipitations (Katasonova & Fedotov, 2009; Salbu, 2009).

4.3 Field-Flow Fractionation

Several recently published reviews emphasize the efficiency of the Field-Flow Fractionation (FFF) techniques for separating and estimating physical parameters of different materials: biopolymers, biological cells, microorganisms, and colloidal and solid particles (Bouby & Geckeis, 2011; Dubascoux et al., 2010; Kowalkowski et al., 2006; Qureshi & Kok, 2011; Stolpe et al., 2005; Williams et al., 2011).

Particles separation by means of the FFF techniques is achieved by a combined action of the non-uniform flow velocity profile of a carrier liquid and a transverse physical field applied perpendicularly to this carrier. Carrier liquid flowing along the channel forms a nearly parabolic flow velocity profile across the channel. The sample to be investigated is dissolved or suspended in a carrier fluid and is pumped through a thin, not filled, channel. At the present time, FFF comprises a family of separation devices with a great number of sub-techniques used mainly for the separation and characterization of particulate species in the size range from 10^{-3} μm to 10^2 μm. Highly popular among those sub-techniques is the flow field-flow fractionation (Fl-FFF). The version of the manufactured separation system for which the channel conveying the carrier liquid had an asymmetric shape (As-Fl-FFF) proved to be the most efficient, as it enabled nanoparticles ranging form 1 nm to 100 μm to be separated. As a general rule, an Inductively Coupled Plasma Mass Spectrometry (ICP-MS) is used as on-line detector, which allows reaching low detection limits, high sensitivity, large dynamic range and ability to simultaneously measure a large number of elements (Table 1).

Method	Analyt	Detection method	Details	References
As-Fl-FFF	Cs, Eu, Th, U	ICP-MS	Interaction of bentonite colloids with metals in presence of HA	Bouby et al. (2011)
As-Fl-FFF	minor elements	UV detector, ICP-MS	Analysis of colloids released from bentonite and crushed rock	Lahtinen et al. (2010)
As-Fl-FFF	U	UV detector, ICP-MS	U complexation by groundwater dissolved organic C	Ranville et al. (2007)
As-Fl-FFF	Cs, La, Ce, Eu, Th, U	ICP-MS	Quantitative characterization of natural colloids	Bouby et al. (2008)
Fl-FFF	U(VI)	ICP-MS	U(VI) sorption to nanoparticulate hematite	Lesher et al. (2009)
Fl-FFF	^{57}Fe, ^{65}Cu, ^{127}I, ^{184}W, ^{88}Sr, ^{238}U	ICP-MS	Chemical and colloidal analyses of natural seep water	Cizdziel et al. (2008)
As-Fl-FFF		LIBD, ICP-MS	Characterization of aquatic groundwater colloids	Baik et al. (2007)
As-Fl-FFF			Cm(III) complexation behavior of isolated HA and FA derived from Opalinus clay	Claret et al. (2005)
As-Fl-FFF	As, Cd, Sb, Se, Sn, Pb	ICP-MS	Soil leachate	Dubascoux et al. (2008)

Table 1. Some field-flow fractionation procedures and detection method used to charactherize colloid particles.

A large number of recent works contributing to the development of the FFF techniques significantly widened their applicability range in particle size analysis (Ahn et al., 2010; Baalousha et al., 2006; 2005; Dubascoux et al., 2008; Gascoyne, 2009; Isaacson & Bouchard, 2010; Otte et al., 2009; Pifer et al., 2011).

4.4 Passive sampling techniques

For a given element, the overwhelming majority of its species are unstable chemical forms that occur under precarious equilibria. As a general rule, these equilibria are disturbed during the routine operations of collection, transport and storage of the samples, this fact resulting in most cases in erroneous information about the considered system. This is the reason why recently, in the elements speciation analysis, passive sampling techniques have received increasingly large acceptance Vrana et al. (2005).

In the most general meaning, passive sampling is that particular sampling technique which relies on the free transfer of the analyte from the sampled environment to a receiving phase in a sampling device, by the effect of the difference between the chemical potentials that the analyte has in the two environments. The transfer of the analyte from one environment toward the other continues until equilibrium is reached within the system, or until the sampling is stopped by an operator. In the first case, it is said that the passive sampling device operates in an equilibrium regime, while in the second one, it is said that it operates in a kinetic regime. In both situations, the sampling occurs without the involvement of any source of energy other than the indicated difference of chemical potential.

Among the passive sampling techniques, the diffusive gradients in thin films (DGT) technique, introduced by Davison & Zhang (1994), is highly ranked as a consequence of its ability to determine labile species in natural waters, sediments and soils. The DGT technique theoretical background relies on the Fick's first law of diffusion. For aqueous systems determinations a passive sampling device is used, which consists of a plastic piece in the shape of a piston, on which two gel discs and a filtering membrane are installed. The first gel, impregnated with binder material, is used for retaining the analytes. The second gel disc has a pre-determined porosity and its role consists in maintaining a constant concentration during the analyte diffusion between the solution and the binder material. The typical binder material is the Chelex-100 resin, while the material used for the diffusion control is the acrylamide/agarose hydrogel. In the end, the binder gel is eluted and the resulting solution is analyzed by means of ICP-MS French et al. (2005); Garmo et al. (2008; 2006), multi-collector ICP-MS (Malinovsky et al., 2005), thermal ionization mass-spectrometry (TIMS) (Dahlqvist et al., 2005), or directly through the gel analysis by laser-ablation ICP-MS (Pearson et al., 2006; Warnken et al., 2004). A comparative study between the DGT techniques and ultrafiltration has been conducted by Forsberg et al. (2006).

By simultaneously immersing several devices which have diffusion gels of various thicknesses or porosities, there is possible to obtain information about the nature of the complexes which are present in various categories of natural waters, and also about the kinetics of the geochemical reactions in which those complexes are involved (Zhang & Davison, 2000; 2001), including in porewater (Leermakers et al., 2005; Wu et al., 2011). DGT has been successfully utilized in order to monitor the radionuclides migration (Chang et al., 1998; Duquène et al., 2010; Gao et al., 2010; Gregusova & Docekal, 2011; Li et al., 2007; 2009; 2006; Salbu, 2007).

The same class of techniques also encompasses the diffusive equilibration in thin films (DET). In this latter case, the sampling device includes only a single layer of gel. This gel layer is maintained in contact with the environment to be analyzed, until equilibrium is reached between the analyte concentration in the environment, and the corresponding concentration in the gel. The technique is mainly utilized in sedimentary environments (Dočekalová et al., 2002). The total content of the analyte in the gel mirrors the ability of the concerned species to penetrate the gel, being controlled by its dimension. In the case of elements associations with colloidal matter, by combining information provided by DET on the equilibration, with information provided by DGT on the species dynamics, a much more appropriate description is obtained in terms of elements speciation (Fones et al., 2001; Gao et al., 2006; van der Veeken et al., 2008; Vandenhove et al., 2007).

In spite of being quite simple to handle, the DGT/DET techniques require a very cautious approach in terms of interpretation. One must take into account that the elements behaviour is not identical with respect to the binder gels, and that equilibration is largely controlled by the pH and the ionic strength of the solution. In addition, potential artifacts may be introduced during the devices preparation, elution and determination processes.

5. Conclusion

All the energy generation, industrial, medical, or military activities which utilize radioactive substances are producing low and intermediate level, and high-level radioactive waste, which needs to be isolated from the biosphere in order to protect the future generations from the hazards potentially induced by the associated radioactivity. As a function of the radioactivity level and of the half life length of the radioisotopes existing in their inventory, either near surface disposal facilities or deep geological repositories are assigned to the storage of that waste.

It is necessary that a radioactive waste repository location is selected very carefully in terms of its hydrogeological environment, since it is a well-known fact that groundwater is the most important vector involved in the transfer of the contaminants. The radionuclides migration across the geosphere takes place as water-dissolved constituents, but especially bound to the particulate phases carried by groundwater. Generally speaking, elements in a "dissolved" state speciate, i.e. they are distributed among forms defined by electronic states, oxidation states, isotopic compositions, as well as complex or molecular specific structures, while when they occur as particles, it is said that they belong to chemical fractions. A chemical fraction is a group of chemical entities which have common physical (e.g. size), or chemical (e.g. reactivity) properties, that group being operationally outlined by means of an analytical process.

Many authors rightly believe that the part played in the radionuclides transfer by particulate phases in general, but especially by the colloidal matter, is so important, that modeling approaches which do not take it into account are unrealistic. In an aqueous environment, radiocolloids occur as intrinsic-colloids and pseudo-colloids. The first category is specific to several transuranic elements which under certain conditions (e.g., pH, ionic strength), possess the capacity of forming structures with colloidal properties. The second type of colloids forms through the attachment of the radionuclide to particles pre-existing in groundwater. Those particles are spontaneously formed in the aquifer structures, but they might also derive from engineered barriers that are built in order to stop radionuclides from migrating out of the repositories. In this respect, bentonitic clays used as a buffer are the most important source of colloidal particles.

In order to identify and to characterize the particulate phases-radionuclides associations, a set of specific separation and determination methods is resorted to. Among the separation techniques, the filtration/ultrafiltration is the most frequently used, while for the colloidal matter characterization, much more efficient are the techniques belonging to the field-flow fractionation category. The latter have the advantage that they may be hyphenated with quantitative determination techniques, among which the inductively coupled plasma mass spectrometry is the most widely used. Recently, into the radiocolloids analysis domain there additionally included the "diffusive gradients in thin films (DGT)" and "diffusive equilibration in thin films (DET)" techniques, which appear to be outstandingly promising for the study of the radionuclides migration across the environment.

6. Acknowledgements

The present study will be conducted in the framework of the research project MIGRELEMENT (Project No. 32112/2008), financially supported by The Executive Agency for Higher Education, Research, Development and Innovation Funding (UEFISCDI).

7. References

Acero, P., Auqué, L. F., Gimeno, M. J. & Gómez, J. B. (2010). Evaluation of mineral precipitation potential in a spent nuclear fuel repository, *Environ. Earth Sci.* 59: 1613–1628.

Ahn, J. Y., Kim, K. H., Lee, J. Y., Williams, P. S. & Moon, M. H. (2010). Effect of asymmetrical flow field-flow fractionation channel geometry on separation efficiency, *J. Chromatogr. A* 1217(24): 3876–3880.

Akgün, H., Koçkar, M. K. & Aktürk, O. (2006). Evaluation of a compacted bentonite/sand seal for underground waste repository isolation, *Environ. Geol.* 50: 331–337.

Albarran, N., Alonso, Ú. Missana, T., García-Gutiérrez, M. & Mingarrom, M. (2008). Evaluation of geochemical conditions favourable for the colloid-mediated uranium migration in a granite fracture, *e-Terra* 5: 1–8.

Albarran, N., Missana, T., García-Gutiérrez, M., Alonso, U. & Mingarro, M. (2011). Strontium migration in a crystalline medium: effects of the presence of bentonite colloids, *J. Contam. Hydrol.* 122(1-4): 76– 85.

Aleksandrova, O. N., Schulz, M. & Matthies, M. (2010). Natural remediation of surface water systems contaminated with nuclear waste via humic substances in South Ural, *Water, Air, Soil Pollut.* 206: 203–214.

Altmann, S. (2008). 'Geo'chemical research: A key building block for nuclear waste disposal safety cases, *J. Contam. Hydrol.* 102(3-4): 174–179.

Anderson, E. B., Rogozin, Y. M., Smirnova, E. A., Bryzgalova, R. V., Malimonova, S. I., Andreeva, N. R., Shabalev, S. I. & Savonenkov, V. G. (2009). Effect of colloidal component in solutions on adsorption of actinides [Am(III), Pu(IV)] from simulated groundwater by glass and granodiorite, *Radiochemistry* 51: 542–550.

Andersson, P. S., Porcelli, D., Gustafsson, O., Ingri, J. & Wasserburg, G. J. (2001). The importance of colloids for the behavior of uranium isotopes in the low-salinity zone of a stable estuary, *Geochim. Cosmochim. Acta* 65(1): 13–25.

Arcos, D., Grandia, F., Domenech, C., Fernandez, A. M., Villar, M. V., Muurinen, A., Carlsson, T., Sellin, P. & Hernan, P. (2008). Long-term geochemical evolution of the near field repository: Insights from reactive transport modelling and experimental evidences, *J. Contam. Hydrol.* 102(3-4): 196–209.

Baalousha, M., Kammer, F. V. D., Motelica-Heino, M. & Hilal, H. S.and Le Coustumer, P. (2006). Size fractionation and characterization of natural colloids by flow-field flow fractionation coupled to multi-angle laser light scattering, *J. Chromatogr. A* 1104: 272–281.

Baalousha, M., Kammer, F. V. D., Motelica-Heino, M. & Le Coustumer, P. (2005). Natural sample fractionation by FlFFF-MALLS-TEM: Sample stabilization, preparation, pre-concentration and fractionation, *J. Chromatogr. A* 1093: 156–166.

Baik, M. H., Kim, S. S., Lee, J. K., Lee, S. Y., Kim, G. Y. & Yun, S. T. (2010). Sorption of ^{14}C, ^{99}Tc, ^{137}Cs, ^{90}Sr, ^{63}Ni, and ^{241}Am onto a rock and a fracture-filling material from the Wolsong low- and intermediate-level radioactive waste repository, Gyeongju, Korea, *J. Radioanal. Nucl. Chem.* 283: 337–345.

Baik, M. H., Lee, S. Y. & Shon, W. J. (2009). Retention of uranium(VI) by laumontite, a fracture-filling material of granite, *J. Radioanal. Nucl. Chem.* 280: 69–77.

Baik, M.-H., Yun, J.-I., Bouby, M., Hahn, P.-S. & Kim, J.-I. (2007). Characterization of aquatic groundwater colloids by a laser-induced breakdown detection and ICP-MS combined with an asymmetric flow field-flow fractionation, *Korean J. Chem. Eng.* 24: 723–729.

Barros, H. & Abril, J. M. (2005). Constraints in the construction and/or selection of kinetic box models for the uptake of radionuclides and heavy metals by suspended particulate matter, *Ecol. Model.* 185(2-4): 371–385.

Barros, H. & Abril, J. M. (2008). Kinetic box models for the uptake of radionuclides and heavy metals by suspended particulate matter: equivalence between models and its implications, *J. Environ. Radioactivity* 99: 146–158.

Barros, H., Laissaoui, A. & Abril, J. M. (2004). Trends of radionuclide sorption by estuarine sediments. Experimental studies using ^{133}Ba as a tracer, *Sci. Total Environ.* 319(1-3): 253–267.

Batuk, D. N., Shiryaev, A. A., Kalmykov, S. N., Batuk, O. N., Romanchuk, A., Shirshin, E. A. & Zubavichus, Y. V. (2011). Sorption and speciation of uranium on silica colloids, *in* S. N. Kalmykov & M. A. Denecke (eds), *Actinide Nanoparticle Research*, Springer Berlin Heidelberg, pp. 315–332.

Bauer, M. & Blodau, C. (2009). Arsenic distribution in the dissolved, colloidal and particulate size fraction of experimental solutions rich in dissolved organic matter and ferric iron, *Geochim. Cosmochim. Acta* 73(3): 529–542.

Bekhit, H. M., El-Kordy, M. A. & Hassan, A. E. (2009). Contaminant transport in groundwater in the presence of colloids and bacteria: Model development and verification, *J. Contam. Hydrol.* 108(3-4): 152–167.

Blo, G., Contado, C., Fagioli, F. & Dondi, F. (2000). Size-elemental characterization of suspended particle matter by split-flow thin cell fractionation and slurry analysis-electrothermal atomic absorption spectrometry, *Analyst* 125(7): 1335–1339.

Bondareva, L. (2011). The relationship of mineral and geochemical composition to artificial radionuclide partitioning in Yenisei river sediments downstream from Krasnoyarsk, *Environ. Monit. Assess.* Online First: 1–17.

Bondareva, L. G. & Bolsunovskii, A. Y. (2008). Speciation of artificial radionuclides ^{60}Co, ^{137}Cs, ^{152}Eu, and ^{241}Am in bottom sediments of the Yenisei river, *Radiochemistry* 50: 547–552.

Bouby, M. & Geckeis, H. (2011).   Characterization of colloid-borne actinides by flow field-flow fractionation (FlFFF) multidetector analysis (MDA), *in* S. N. Kalmykov & M. A. Denecke (eds), *Actinide Nanoparticle Research*, Springer Berlin Heidelberg, pp. 105–135.

Bouby, M., Geckeis, H. & Geyer, F. W. (2008).   Application of asymmetric flow field-flow fractionation (AsFlFFF) coupled to inductively coupled plasma mass spectrometry (ICPMS) to the quantitative characterization of natural colloids and synthetic nanoparticles, *Anal. Bioanal. Chem.* 392: 1447–1457.

Bouby, M., Geckeis, H., Lutzenkirchen, J., Mihai, S. & Schafer, T. (2011).   Interaction of bentonite colloids with Cs, Eu, Th and U in presence of humic acid: A flow field-flow fractionation study, *Geochim. Cosmochim. Acta* 75(13): 3866–3880.

Bowie, A. R., Townsend, A. T., Lannuzel, D., Remenyi, T. A. & van der Merwe, P. (2010).   Modern sampling and analytical methods for the determination of trace elements in marine particulate material using magnetic sector inductively coupled plasma-mass spectrometry, *Anal. Chim. Acta* 676(1-2): 15–27.

Bradbury, M. H. & Baeyens, B. (2011).   Predictive sorption modelling of Ni(II), Co(II), Eu(III), Th(IV) and U(VI) on MX-80 bentonite and Opalinus Clay: A bottom-up approach, *Appl. Clay Sci.* 52(1-2): 27–33.

Bradbury, M. H., Baeyens, B., Geckeis, H. & Rabung, T. (2005).   Sorption of Eu(III)/Cm(III) on Ca-montmorillonite and Na-illite. Part 2: Surface complexation modelling, *Geochim. Cosmochim. Acta* 69(23): 5403–5412.

Bradford, S. A. & Bettahar, M. (2006).   Concentration dependent transport of colloids in saturated porous media, *J. Contam. Hydrol.* 82(1-2): 99–117.

Bryan, N. D., Barlow, J., Warwick, P., Stephens, S., Higgo, J. J. W. & Griffin, D. (2005).   The simultaneous modelling of metal ion and humic substance transport in column experiments, *J. Environ. Monit.* 7: 196–202.

Buesseler, K. O., Kaplan, D. I., Dai, M. & Pike, S. (2009).   Source-dependent and source-independent controls on plutonium oxidation state and colloid associations in groundwater, *Environ. Sci. Technol.* 43(5): 1322–1328.

Butler, B. A., Ranville, J. F. & Ross, P. E. (2008).   Direct versus indirect determination of suspended sediment associated metals in a mining-influenced watershed, *Appl. Geochem.* 23: 1218–1231.

Cadini, F., De Sanctis, J., Girotti, T., Zio, E., Luce, E. & Taglioni, A. (2010).   Monte Carlo estimation of radionuclide release at a repository scale, *Ann. Nucl. Energy* 37(6): 861–866.

Caron, F. & Smith, D. (2011).   Fluorescence analysis of natural organic matter fractionated by ultrafiltration: Contrasting between urban-impacted water, and radio-contaminated water from a near-pristine site, *Water, Air, Soil Pollut.* 214: 471–490.

Chang, L.-Y., Davison, W., Zhang, H. & Kelly, M. (1998).   Performance characteristics for the measurement of Cs and Sr by diffusive gradients in thin films (DGT), *Anal. Chim. Acta* 368(3): 243 – 253.

Cidu, R. & Frau, F. (2009).   Distribution of trace elements in filtered and non filtered aqueous fractions: Insights from rivers and streams of Sardinia (Italy), *Appl. Geochem.* 24(4): 611–623.

Cizdziel, J. V., Guo, C., Steinberg, S. M., Yu, Z. & Johannesson, K. H. (2008). Chemical and colloidal analyses of natural seep water collected from the exploratory studies facility inside Yucca Mountain, Nevada, USA, *Environ. Geochem. Health* 30(1): 31–44.

Claret, F., Schäfer, T., Rabung, T., Wolf, M., Bauer, A. & Buckau, G. (2005). Differences in properties and Cm(III) complexation behavior of isolated humic and fulvic acid derived from Opalinus clay and Callovo-Oxfordian argillite, *Appl. Geochem.* 20(6): 1158–1168.

Constantinou, E. & Pashalidis, I. (2010). Uranium determination in water samples by liquid scintillation counting after cloud point extraction, *J. Radioanal. Nucl. Chem.* 286: 461–465.

Constantinou, E. & Pashalidis, I. (2011). Thorium determination in water samples by liquid scintillation counting after its separation by cloud point extraction, *J. Radioanal. Nucl. Chem.* 287: 261–265.

Cortecci, G., Boschetti, T., Dinelli, E., Cidu, R., Podda, F. & Doveri, M. (2009). Geochemistry of trace elements in surface waters of the Arno River Basin, northern Tuscany, Italy, *Appl. Geochem.* 24(5): 1005–1022.

Curti, E., Dähn, R., Farges, F. & Vespa, M. (2009). Na, Mg, Ni and Cs distribution and speciation after long-term alteration of a simulated nuclear waste glass: A micro-XAS/XRF/XRD and wet chemical study, *Geochim. Cosmochim. Acta* 73(8): 2283–2298.

Dahlqvist, R., Andersson, P. S. & Ingri, J. (2005). The concentration and isotopic composition of diffusible Nd in fresh and marine waters, *Earth Planet. Sci. Lett.* 233(1-2): 9–16.

Davison, W. & Zhang, H. (1994). *In-situ* speciation measurements of trace components in natural waters using thin-film gels, *Nature* 367(6463): 546–548.

Davydov , Y. P., Voronik, N. I., Davydov, D. Y. & Titov, A. S. (2006). Speciation of Cr(III) radionuclides in solutions, *Radiochemstry* 48(4): 365–368.

Davydov, Y. P., Davydov, D. Y. & Zemskova, L. M. (2006). Speciation of Zr(IV) radionuclides in solutions, *Radiochemstry* 48(4): 358–364.

Davydov, Y. P., Grachok, M. A. & Davydov, D. Y. (2003). Speciation of Fe(III) radionuclides in aqueous solutions, *Radiochemistry* 45: 40–46.

De Windt, L., Pellegrini, D. & van der Lee, J. (2004). Coupled modeling of cement/claystone interactions and radionuclide migration, *J. Contam. Hydrol.* 68(3-4): 165–182.

Deepthi Rani, R. & Sasidhar, P. (2011). Stability assessment and characterization of colloids in coastal groundwater aquifer system at Kalpakkam, *Environ. Earth Sci.* 62: 233–243.

Degueldre, C. & Bolek, M. (2009). Modelling colloid association with plutonium: The effect of pH and redox potential, *Appl. Geochem.* 24(2): 310–318.

Del Nero, M., A., A., Madé, B., Barillon, R. & Duplâtre, G. (2004). Surface charges and Np(V) sorption on amorphous al and fe silicates, *Chem. Geol.* 211(1-2): 15–45.

Delos, A., Walther, C., Schäfer, T. & Büchner, S. (2008). Size dispersion and colloid mediated radionuclide transport in a synthetic porous media, *J. Colloid Interface Sci.* 324(1-2): 212–215.

Dogaru, D., Niculae, O., Terente, M., Jinescu, G. & Duliu, O. G. (2010). Complementary safety indicators of Saligny radioactive waste repository, *Romanian Reports Phys.* 62(4): 811–820.

Doucet, F. D., Maguire, L. & Lead, J. R. (2005). Assessment of cross-flow filtration for the size fractionation of freshwater colloids and particles, *Talanta* 67: 144–154.

Doucet, F. J., Maguire, L. & Lead, J. R. (2004). Size fractionation of aquatic colloids and particles by cross-flow filtration: analysis by scanning electron and atomic force microscopy, *Anal. Chim. Acta* 522: 59–71.

Dočekalová, H., Clarisse, O., Salomon, S. & Wartel, M. (2002). Use of constrained DET probe for a high-resolution determination of metals and anions distribution in the sediment pore water, *Talanta* 57(1): 145–155.

Dreissig, I., Weiss, S., Hennig, c., Bernhard, G. & Zänker, H. (2011). Formation of uranium(IV)-silica colloids at near-neutral pH, *Geochim. Cosmochim. Acta* 75(2): 352–367.

Dubascoux , S., Von Der Kammer, F., Le Hécho, I., Gautier, M. P. & Lespes, G. (2008). Optimisation of asymmetrical flow field flow fractionation for environmental nanoparticles separation, *J. Chromatogr. A* 1206: 160–165.

Dubascoux, S., Le Hécho, I., Hassellöv, M., Von Der Kammer, F., Gautier, M. P. & Lespes, G. (2010). Field-flow fractionation and inductively coupled plasma mass spectrometer coupling: History, development and applications, *J. Anal. At. Spectrom.* 25(5): 613–623.

Dubascoux, S., Le Hécho, I., Gautier, M. P. & Lespes, G. (2008). On-line and off-line quantification of trace elements associated to colloids by As-Fl-FFF and ICP-MS, *Talanta* 77: 60–65.

Duquène, L., Vandenhove, H., Tack, F., Van Hees, M. & Wannijn, J. (2010). Diffusive gradient in thin FILMS (DGT) compared with soil solution and labile uranium fraction for predicting uranium bioavailability to ryegrass, *J. Environ. Radioactivity* 101(2): 140–147.

Eyrolle, F. & Charmasson, S. (2004). Importance of colloids in the transport within the dissolved phase (< 450 nm) of artificial radionuclides from the Rhône river towards the Gulf of Lions (Mediterranean Sea), *J. Environ. Radioact.* 72(3): 273–286.

Ferrage, E., Tournassat, C., Rinnert, E. & Lanson, B. (2005). Influence of pH on the interlayer cationic composition and hydration state of Ca-montmorillonite: Analytical chemistry, chemical modelling and XRD profile modelling study, *Geochim. Cosmochim. Acta* 69(11): 2797–2812.

Filby, A., Plaschke, M., Geckeis, H. & Fanghánel, T. (2008). Interaction of latex colloids with mineral surfaces and Grimsel granodiorite, *Journal of Contaminant Hydrology* 102(3-4): 273–284.

Fones, G. R., Davison, W., Holby, O., Jorgensen, B. B. & Thamdrup, B. (2001). High-resolution metal gradients measured by in situ DGT/DET deployment in Black Sea sediments using an autonomous benthic lander, *Limnol. Oceanogr.* 46(4): 982–988.

Forsberg, J., Dahlqvist, R., Gelting-Nyström, J. & Ingri, J. (2006). Trace metal speciation in brackish water using diffusive gradients in thin films and ultrafiltration: Comparison of techniques, *Environ. Sci. Technol.* 40(12): 3901–3905.

French, M. A., Zhang, H., Pates, J. M., Bryan, S. E. & Wilson, R. C. (2005). Development and performance of the diffusive gradients in thin-films technique for the measurement of technetium-99 in seawater, *Anal. Chem.* 77(1): 135–139.

Froideval, A., Del Nero, M., Gaillard, C., Barillon, R., Rossini, I. & Hazemann, J. (2006). Uranyl sorption species at low coverage on Al-hydroxide: TRLFS and XAFS studies, *Geochim. Cosmochim. Acta* 70(21): 5270–5284.

Galamboš, M., Kufčáková, J. & Rajec, P. (2009). Adsorption of cesium on domestic bentonites, *J. Radioanal. Nucl. Chem.* 281: 485–492.

Galamboš, M., Rosskopfová, O., Kufčáková, J. & Rajec, P. (2011). Utilization of Slovak bentonites in deposition of high-level radioactive waste and spent nuclear fuel, *J. Radioanal. Nucl. Chem.* 288: 765–777.

Gammons, C. H., Nimick, D. A., Parker, S. R., Cleasby, T. E. & McCleskey, R. B. (2005). Diel behavior of iron and other heavy metals in a mountain stream with acidic to neutral pH: Fisher Creek, Montana, USA, *Geochim. Cosmochim. Acta* 69: 2505–2516.

Gao, Y., Baeyens, W., Galan, S. D., Poffijn, A. & Leermakers, M. (2010). Mobility of radium and trace metals in sediments of the Winterbeek: Application of sequential extraction and DGT techniques, *Environ. Poll.* 158(7): 2439–2445.

Gao, Y., Leermakers, M., Gabelle, C., Divis, P., Billon, G., Ouddane, B., Fischer, J.-C., Wartel, M. & Baeyens, W. (2006). High-resolution profiles of trace metals in the pore waters of riverine sediment assessed by DET and DGT, *Sci. Total Environ.* 362(1-3): 266–277.

Garmo, Ø. A., Davison, W. & Zhang, H. (2008). Effects of binding of metals to the hydrogel and filter membrane on the accuracy of the diffusive gradients in thin films technique, *Anal. Chem.* 80(23): 9220–9225.

Garmo, Ø. A., Lehto, N. J., Zhang, H., Davison, W., Røyset, O. & Steinnes, E. (2006). Dynamic aspects of dgt as demonstrated by experiments with lanthanide complexes of a multidentate ligand, *Environ. Sci. Technol.* 40(15): 4754–4760.

Gascoyne, P. R. C. (2009). Dielectrophoretic-field flow fractionation analysis of dielectric, density, and deformability characteristics of cells and particles, *Anal. Chem.* 81(21): 8878–8885.

Gaucher, E. C., Blanc, P., Matray, J.-M. & Michau, N. (2004). Modeling diffusion of an alkaline plume in a clay barrier, *Appl. Geochem.* 19(10): 1505–1515.

Geckeis, H. & Rabung, T. (2008). Actinide geochemistry: From the molecular level to the real system, *J. Contam. Hydrol.* 102(3-4): 187–195.

Geckeis, H., Rabung, T. & Schäfer, T. (2011). Actinide-nanoparticle interaction: Generation, stability and mobility, *in* S. N. Kalmykov & M. A. Denecke (eds), *Actinide Nanoparticle Research*, Springer Berlin Heidelberg, pp. 1–30.

Geraedts, K. & Maes, A. (2008). Determination of the conditional interaction constant between colloidal technetium(IV) and Gorleben humic substances, *Appl. Geochem.* 23: 1127–1139.

Gimbert, L., Haygarth, P., Beckett, R. & Worsfold, P. (2005). Comparison of centrifugation and filtration techniques for the size fractionation of colloidal material in soil suspensions using sedimentation field-flow fractionation, *Environ. Sci. Technol.* 39(6): 1731–1735.

Goryachenkova, T. A., Kazinskaya, I. E., Kuzovkina, E. V.and Novikov, A. P. & Myasoedov, B. F. (2009). Association of radionuclides with colloids in soil solutions, *Radiochemistry* 51: 201–210.

Goveia, D., Lobo, F. A., Burba, P., Fraceto, L., Dias Filho, N. L. & Rosa, A. H. (2010). Approach combining on-line metal exchange and tangential-flow ultrafiltration for in-situ characterization of metal species in humic hydrocolloids, *Anal. Bioanal. Chem.* 397(2): 851–860.

Graham, M. C., Oliver, I. W., MacKenzie, A. B., Ellam, R. M. & Farmer, J. G. (2008). An integrated colloid fractionation approach applied to the characterisation of porewater uranium-humic interactions at a depleted uranium contaminated site, *Sci. Total Environ.* 404(1): 207–217.

Graham, M. C., Oliver, I. W., MacKenzie, A. B., Ellam, R. M. & Farmer, J. G. (2011). Mechanisms controlling lateral and vertical porewater migration of depleted uranium (DU) at two UK weapons testing sites, *Sci. Total Environ.* 409(10): 1854–1866.

Grambow, B. (2008). Mobile fission and activation products in nuclear waste disposal, *J. Contam. Hydrol.* 102(3-4): 180–186.

Gregusova, M. & Docekal, B. (2011). New resin gel for uranium determination by diffusive gradient in thin films technique, *Anal. Chim. Acta* 684(1-2): 142–146.

Grolimund, D., Barmettler, K. & Borkovec, M. (2001). Release and transport of colloidal particles in natural porous media 2. Experimental results and effects of ligands, *Water Resour. Res.* 37(3): 571–582.

Grolimund, D. & Borkovec, M. (2001). Release and transport of colloidal particles in natural porous media 1. Modeling, *Water Resour. Res.* 37(3): 559–570.

Grolimund, D. & Borkovec, M. (2006). Release of colloidal particles in natural porous media by monovalent and divalent cations, *J. Contam. Hydrol.* 87(3-4): 155–175.

Gustafsson, O. & Gschwend, P. M. (1997). Aquatic colloids: Concepts, definitions, and current challenges, *Limnol. Oceanogr.* 42(3): 519–528.

Hassellöv, M., Buesseler, K. O., Pike, S. M. & Dai, M. (2007). Application of cross-flow ultrafiltration for the determination of colloidal abundances in suboxic ferrous-rich ground waters, *Sci. Total Environ.* 372(2-3): 636–644.

Hedberg, Y., Herting, G. & Wallinder, I. O. (2011). Risks of using membrane filtration for trace metal analysis and assessing the dissolved metal fraction of aqueous media - A study on zinc, copper and nickel, *Environ. Poll.* 159(5): 1144–1150.

Hu , Q.-H., Weng, J.-Q. & Wang, J.-S. (2010). Sources of anthropogenic radionuclides in the environment: a review, *J. Environ. Radioact.* 101(6): 426–437.

Hu, B., Cheng, W., Zhang, H. & Yang, S. (2010). Solution chemistry effects on sorption behavior of radionuclide 63ni(ii) in illite-water suspensions, *J. Nucl. Mater.* 406(2): 263–270.

Hu, J., Xie, Z., He, B., Sheng, G., Chen, C., Li, J., Chen, Y. & Wang, X. (2010). Sorption of Eu(III) on GMZ bentonite in the absence/presence of humic acid studied by batch and XAFS techniques, *Sci. China, Chemistry* 53: 1420–1428.

Hu, Q. & Mori, A. (2008). Radionuclide transport in fractured granite interface zones, *Phys. Chem. Earth, Pts. A/B/C* 33(14-16): 1042–1049.

IAEA (2003). Radioactive waste management glossary, *Technical Report IAEAL 03-00320*, International Atomic Energy Agency, Vienna.

IAEA (2004). Surveillance and monitoring of near surface disposal facilities for radioactive waste, *Technical Report Safety reports series, No. 35*, International Atomic Energy Agency, Vienna.

IAEA (2007). IAEA safety glossary. Terminology used in nuclear safety and radiation protection, *Technical Report IAEAL 07-00481*, International Atomic Energy Agency, Vienna.

Ilina, T., Panfilov, M., Buès, M. & Panfilova, I. (2008). A pseudo two-phase model of colloid transport in porous media, *Transp. Porous Med.* 71: 311–329.

Isaacson, C. W. & Bouchard, D. (2010). Asymmetric flow field flow fractionation of aqueous $C_6 0$ nanoparticles with size determination by dynamic light scattering and quantification by liquid chromatography atmospheric pressure photo-ionization mass spectrometry, *J. Chromatogr. A* 1217(9): 1506–1512.

Jeong, M.-S., Hwang, Y. & Kang, C.-H. (2011). Modeling of the pseudo-colloids migration for multi-member decay chains with a arbitrary flux boundary condition in a fractured porous medium, *J. Radioanal. Nucl. Chem.* 289: 287–293.

Johnsson, A., Ödegaard Jensen, A., Jakobsson, A.-M., Ekberg, C. & Pedersen, K. (2008). Bioligand-mediated partitioning of radionuclides to the aqueous phase, *J. Radioanal. Nucl. Chem.* 277: 637–644.

Joseph, C., Schmeide, K., Sachs, S., Brendler, V., Geipel, G. & Bernhard, G. (2011). Sorption of uranium(VI) onto Opalinus Clay in the absence and presence of humic acid in Opalinus Clay pore water, *Chem. Geol.* 284(3-4): 240–250.

Kalmykov, S. N. & Denecke, M. A. (eds) (2011). *Actinide Nanoparticle Research*, Springer-Verlag, Berlin.

Kalmykov, S. N., Zakharova, E. V., Novikov, A. P., Myasoedov, B. F. & Utsunomiya, S. (2011). Effect of redox conditions on actinide speciation and partitioning with colloidal matter, *in* S. N. Kalmykov & M. A. Denecke (eds), *Actinide Nanoparticle Research*, Springer Berlin Heidelberg, pp. 361–375.

Katasonova, O. & Fedotov, P. (2009). Methods for continuous flow fractionation of microparticles: Outlooks and fields of application, *J. Anal. Chem.* 64(3): 212–225.

Kelkar, S., Ding, M., Chu, S., Robinson, B. A., Arnold, B., Meijer, A. & Eddebbarh, A.-A. (2010). Modeling solute transport through saturated zone ground water at 10 km scale: Example from the Yucca Mountain license application, *J. Contam. Hydrol.* 117(1-4): 7–25.

Kersting, A. B. & Zavarin, M. (2011). Colloid-facilitated transport of plutonium at the Nevada Test Site, NV, USA, *in* S. N. Kalmykov & M. A. Denecke (eds), *Actinide Nanoparticle Research*, Springer Berlin Heidelberg, pp. 399–412.

Khasanova, A. B., Shcherbina, N. S., Kalmykov, S. N., Teterin, Y. A. & Novikov, A. P. (2007). Sorption of Np(V), Pu(V), and Pu(IV) on colloids of Fe(III) oxides and hydrous oxides and MnO_2, *Radiochemistry* 49: 419–425.

Kiliari, T. & Pashalidis, I. (2010). The effect of physicochemical parameters on the separation recovery of plutonium and uranium from aqueous solutions by cation exchange, *J. Radioanal. Nucl. Chem.* 286: 467–470.

Kim, J.-S., Kwon, S.-K., Sanchez, M. & Cho, G.-C. (2011). Geological storage of high level nuclear waste, *KSCE J. Civil Eng.* 15: 721–737.

Kosakowski, G. (2004). Anomalous transport of colloids and solutes in a shear zone, *J. Contam. Hydrol.* 72(1-4): 23–46.

Kowalkowski, T., Buszewski, B., Cantado, C. & Dondi, F. (2006). Field-flow fractionation: Theory, techniques, applications and the challenges, *Crit. Rev. Anal. Chem.* 36(2): 129–135.

Kretzschmar, R., Barmettler, K., Grolimund, D., Yan, Y., Borkovec, M. & Sticher, H. (1997). Experimental determination of colloid deposition rates and collision efficiencies in natural porous media, *Water Resour. Res.* 33(5): 1129–1137.

Ku, T. L., Luo, S., Goldstein, S. J., Murrell, M. T., Chu, W. L. & Dobson, P. F. (2009). Modeling non-steady state radioisotope transport in the vadose zone - A case study using uranium isotopes at Peña Blanca, Mexico, *Geochim. Cosmochim. Acta* 73(20): 6052–6064.

Kulyako, Y. M., Mal'kovskii, V. I., Trofimov, T. I., Myasoedov, B. F., Fujiwara, A. & Tochiyama, O. (2008). Formation of polymeric Pu(IV) hydroxide structures in aqueous solutions, *Radiochemistry* 50: 594–600.

Kunze, P., Seher, H., Hauser, W., Panak, P. J., Geckeis, H., Fanghänel, T. & Schäfer, T. (2008). The influence of colloid formation in a granite groundwater bentonite porewater mixing zone on radionuclide speciation, *J. Contam. Hydrol.* 102(3-4): 263–272.

Kurosawa, S. & Ueta, S. (2001). Effect of colloids on radionuclide migration for performance assessment of HLW disposal in Japan, *Pure Appl. Chem.* 73(12): 2027–2037.

Kyriakou, M. & Pashalidis, I. (2011). Application of different types of resins in the radiometric determination of uranium in waters, *J. Radioanal. Nucl. Chem.* 287: 773–778.

Lahtinen, M., Hölttä, P., Riekkola, M. L. & Yohannes, G. (2010). Analysis of colloids released from bentonite and crushed rock, *Phys. Chem. Earth, Pts. A/B/C* 35: 265–270.

Laverov, N. P., Velichkin, V. I., Malkovsky, V. I., Tarasov, N. N. & Dikov, Y. P. (2010). A comprehensive study of the spread of radioactive contamination in the geological medium near Lake Karachai, Chelyabinsk oblast, *Geol. Ore Deposits* 52: 5–13.

Leermakers, M., Gao, Y., Gabelle, C., Lojen, S., Ouddane, B., Wartel, M. & Baeyens, W. (2005). Determination of high resolution pore water profiles of trace metals in sediments of the Rupel River (Belgium) using DET (Diffusive Equilibrium in Thin Films) and DGT (Diffusive Gradients in Thin Films) techniques, *Water Air Soil Pollut.* 166(1–4): 265–286.

Lesher, E. K., Ranville, J. F. & Honeyman, B. D. (2009). Analysis of pH dependent uranium(VI) sorption to nanoparticulate hematite by flow field-flow fractionation - inductively coupled plasma mass spectrometry, *Environ. Sci. Technol.* 43(14): 5403–5409.

Li, W., Li, C., Zhao, J. & Cornett, R. J. (2007). Diffusive gradients in thin films technique for uranium measurements in river water, *Anal. Chim. Acta* 592(1): 106 – 113.

Li, W., Wang, F., Zhang, W. & Evans, D. (2009). Measurement of stable and radioactive cesium in natural waters by the diffusive gradients in thin films technique with new selective binding phases, *Anal. Chem.* 81(14): 5889–5895.

Li, W., Zhao, J., Li, C., Kiser, S. & Cornett, R. J. (2006). Speciation measurements of uranium in alkaline waters using diffusive gradients in thin films technique, *Anal. Chim. Acta* 575(2): 274–280.

Li, Y., Tian, S. & Qian, T. (2011). Transport and retention of strontium in surface-modified quartz sand with different wettability, *J. Radioanal. Nucl. Chem.* 289: 337–343.

Li, Z., Zhang, D. & Li, X. (2010). Tracking colloid transport in porous media using discrete flow fields and sensitivity of simulated colloid deposition to space discretization, *Environ. Sci. Technol.* 44(4): 1274–1280.

Lippold , H., Mansel, A. & Kupsch, H. (2005). Influence of trivalent electrolytes on the humic colloid-borne transport of contaminant metals: competition and flocculation effects, *J. Contam. Hydrol.* 76(3-4): 337–352.

Lippold, H., Müller, N. & Kupsch, H. (2005). Effect of humic acid on the pH-dependent adsorption of terbium (III) onto geological materials, *Appl. Geochem.* 20(6): 1209–1217.

Liu, J., Andya, J. D. & Shire, S. J. (2006). A critical review of analytical ultracentrifugation and field flow fractionation methods for measuring protein aggregation, *AAPS J.* 8: E580–E589.

Loux, N. T. (2011). Simulating the stability of colloidal amorphous iron oxide in natural water, *Water, Air, Soil Pollut.* 217: 157–172.

Lujanienė, G., Beneš, P., Štamberg, K., Šapolaitė J., Vopalka, D., Radžiūtė, E. & Ščiglo, T. (2010). Effect of natural clay components on sorption of Cs, Pu and Am by the clay, *J. Radioanal. Nucl. Chem.* 286: 353–359.

Luk'yanova, E. A., Zakharova, E. V., Konstantinova, L. I. & Nazina, T. N. (2008). Sorption of radionuclides by microorganisms from a deep repository of liquid low-level waste, *Radiochemistry* 50: 85–90.

Maity, S., Mishra, S., Bhalke, S., Pandit, G., Puranik, V. & Kushwaha, H. (2011). Estimation of distribution coefficient of polonium in geological matrices around uranium mining site, *J. Radioanal. Nucl. Chem.* 290(1): 75–79.

Malinovsky, D., Dahlqvist, R., Baxter, D. C., Ingri, J. & Rodushkin, I. (2005). Performance of diffusive gradients in thin films for measurement of the isotopic composition of soluble Zn, *Anal. Chim. Acta* 537(1-2): 401 – 405.

Mal'kovskii, V. I. (2011). Deposition of heterogeneous radiocolloid from groundwater on enclosing rocks, *Dokl. Earth Sci.* 436: 39–41.

Malkovsky, V. (2011). Theoretical analysis of colloid-facilitated transport of radionuclides by groundwater, *in* S. N. Kalmykov & M. A. Denecke (eds), *Actinide Nanoparticle Research*, Springer Berlin Heidelberg, pp. 195–243.

Malkovsky, V. I., Dikov, Y. D., Kalmykov, S. N. & Buleev, M. I. (2009). Structure of colloid particles in groundwaters on the territory of the Mayak production association and its impact on the colloid transport of radionuclides in subsoil environments, *Geochem. Int.* 47: 1100–1106.

Malkovsky, V. I. & Pek, A. A. (2009a). Effect of colloids on transfer of radionuclides by subsurface water, *Geol. Ore Deposits* 51: 79–92.

Malkovsky, V. I. & Pek, A. A. (2009b). Effect of elevated velocity of particles in groundwater flow and its role in colloid-facilitated transport of radionuclides in underground medium, *Transp. Porous Med.* 78: 277–294.

Mansur, M. S. & Mushtaq, A. (2011). Separation of yttrium-90 from strontium-90 via colloid formation, *J. Radioanal. Nucl. Chem.* 288: 337–340.

Massoudieh, A. & Ginn, T. R. (2007). Modeling colloid-facilitated transport of multi-species contaminants in unsaturated porous media, *J. Contam. Hydrol.* 92(3-4): 162–183.

Matisoff, G., Ketterer, M. E., Rosen, K., Mietelski, J. W., Vitko, L. F., Persson, H. & Lokas, E. (2011). Downward migration of Chernobyl-derived radionuclides in soils in Poland and Sweden, *Appl. Geochem.* 26(1): 105–115.

Matsunaga, T., Nagao, S., Ueno, T., Takeda, S., Amano, H. & Tkachenko, Y. (2004). Association of dissolved radionuclides released by the Chernobyl accident with colloidal materials in surface water, *Appl. Geochem.* 19(10): 1581–1599.

May, C. C., Worsfold, P. J. & Keith-Roach, M. J. (2008). Analytical techniques for speciation analysis of aqueous long-lived radionuclides in environmental matrices, *Trends Anal. Chemi.* 27(2): 160–168.

Mazurek, M., Alt-Epping, P., Bath, A., Gimmi, T., Waber, H. N., Buschaert, S., De Cannière, P., De Craen, M., Gautschi, A., Savoye, S., Vinsot, A., Wemaere, I. & Wouters, L. (2011). Natural tracer profiles across argillaceous formations, *Appl. Geochem.* 26(7): 1035–1064.

Mibus, J., Sachs, S., Pfingsten, W., Nebelung, C. & Bernhard, G. (2007). Migration of uranium(IV)/(VI) in the presence of humic acids in quartz sand: A laboratory column study, *J. Contam. Hydrol.* 89(3-4): 199–217.

Missana, T., Alonso, U., García-Gutiérrez, M. & Mingarro, M. (2008). Role of bentonite colloids on europium and plutonium migration in a granite fracture, *Appl. Geochem.* 23(6): 1484–1497.

Missana, T. & García-Gutiérrez, M. (2007). Adsorption of bivalent ions (Ca(II), Sr(II) and Co(II)) onto FEBEX bentonite, *Phys. Chem. Earth* 32: 559–567.

Monte, L. (2010). Modelling multiple dispersion of radionuclides through the environment, *J. Environ. Radioact.* 101(2): 134–139.

Monte, L., Periañez, R., Boyer, P., Smith, J. T. & Brittain, J. E. (2009). The role of physical processes controlling the behaviour of radionuclide contaminants in the aquatic environment: a review of state-of-the-art modelling approaches, *J. Environ. Radioact.* 100(9): 779–784.

Morel, F. M. M. (1983). *Principles of Aquatic Chemistry*, John Wiley and Sons, New York.

Murakami, T., Sato, T., Ohnuki, T. & Isobe, H. (2005). Field evidence for uranium nanocrystallization and its implications for uranium transport, *Chem. Geol.* 221(1-2): 117–126.

Niculae, O., Andrei, V., Ionita, G. & Duliu, O. G. (2009). Preliminary safety concept for disposal of the very low level radioactive waste in Romania, *Appl. Radiation Isotopes* 67: 935–938.

Nordstrom, D. K., Alpers, C. N., Coston, J. A., Taylor, H. E., McCleskey, R. B., Ball, J. W., Ogle, S., Cotsifas, J. S. & Davis, J. A. (1999). Geochemistry, toxicity, and sorption properties of contaminated sediments and pore waters from two reservoirs receiving acid mine drainage, *U.S. Geological Survey Toxic Substances Hydrology Program. Proceedings Of The Technical Meeting, Charleston, South Carolina, March 8-12, 1999. U.S. Geological Survey Water-Resources Investigations Report 99-4018a. Volume 1. Contamination From Hard-Rock Mining*, Vol. 1, pp. 289–296.

Novikov , A. P., Kalmykov, S. N., Kuzovkina, E. V., Myasoedov, B. F., Fujiwara, K. & Fujiwara, A. (2009). Evolution of actinide partitioning with colloidal matter collected at "Mayak" site as studied by sequential extraction, *J. Radioanal. Nucl. Chem.* 280: 629–634.

Novikov, A. P., Kalmykov, S. N., Goryachenkova, T. A., Kazinskaya, I. E., Barsukova, K. V., Lavrinovich, E. A., Kuzovkina, E. V. & Myasoedov, B. F. (2009). Association of radionuclides with the colloidal matter of underground waters taken from observation wells in the zone of impact of Lake Karachai, *Radiochemistry* 51: 644–648.

Novikow, A., Goryachenkova, T. A., Kalmykov, S. N., Vlasova, I. E., Kazinskaya, I. E., Barsukova, K. V., Lavrinovich, E. A., Kuzovkina, E. V., Tkachev, V. V. & Myasoedov, B. F. (2009). Speciation of radionuclides in colloidal matter of underground waters taken from observation wells in the zone of impact of Lake Karachai, *Radiochemistry* 51: 649–653.

Ödman, F., Ruth, T. & Pontér, C. (1999). Validation of a field filtration technique for characterization of suspended particulate matter from freshwater. Part I. Major elements, *Appl. Geochem.* 14: 301–317.

Ödman, F., Ruth, T., Rodushkin, I. & Pontér, C. (2006). Validation of a field filtration technique for characterization of suspended particulate matter from freshwater. Part II. Minor, trace and ultra trace elements, *Appl. Geochem.* 21(12): 2112–2134.

Ohtsuka, Y., Yamamoto, M., Takaku, Y., Hisamatsu, S. & Inaba, J. (2006). Cascade ultrafiltering of ^{210}Pb and ^{210}Po in freshwater using a tangential flow filtering system, *J. Radioanal. Nucl. Chem.* 268: 397–403.

Ollivier, P., Radakovitch, O. & Hamelin, B. (2011). Major and trace element partition and fluxes in the Rhône River, *Chem. Geol.* 285(1-4): 15–31.

Otosaka, S., Amano, H., Ito, T., Kawamura, H., Kobayashi, T., Suzuki, T., Togawa, O., Chaykovskaya, E. L., Lishavskaya, T. S., Novichkov, V. P., Karasev, E. V., Tkalin, A. V. & Volkov, Y. N. (2006). Anthropogenic radionuclides in sediment in the Japan Sea: distribution and transport processes of particulate radionuclides, *J. Environ. Radioact.* 91(3): 128–145.

Otte, T., Brüll, R., Macko, T., Pasch, H. & Klein, T. (2009). Optimisation of ambient and high temperature asymmetric flow field-flow fractionation with dual/multi-angle light scattering and infrared/refractive index detection, *J. Chromatogr. A* 1217(5): 722–730.

Panfilov, M., Panfilova, I. & Stepanyants, Y. (2008). Mechanisms of particle transport acceleration in porous media, *Transp. Porous Med.* 74: 49–71.

Pathak, P. N. & Choppin, G. R. (2006). Kinetic and thermodynamic studies of cesium(I) sorption on hydrous silica, *J. Radioanal. Nucl. Chem.* 270: 299–305.

Pathak, P. N. & Choppin, G. R. (2007). Sorption of Am^{3+} cations on suspended silicate: Effects of pH, ionic strength, complexing anions, humic acid and metal ions, *J. Radioanal. Nucl. Chem.* 274: 517–523.

Pearson, D., Nowell, G., Widerlund, A. & Davison, W. (2006). Quantifying micro-niche behaviour in sediments: Precise and accurate measurements of DGT gels by laser ablation high resolution laser ablation ICPMS, *Geochim. Cosmochim. Acta* 70(18, Supplement 1): A478–A478.

Perevalov, S. A., Kulyako, Y. M., Vinokurov, S. E., Myasoedov, B. F., Fujiwara, A. & Tochiyama, O. (2009). Sorption of Pu(IV) in the polymeric colloidal form on rock typical of Mayak Production Association area, *Radiochemistry* 51: 373–377.

Pérez del Villar, L., Delgado, A., Reyes, E., Pelayo, M., Fernández-Soler, J. M., Cózar, J. S., Tsige, M. & Quejido, A. J. (2005). Thermochemically induced transformations in Al-smectites: A Spanish natural analogue of the bentonite barrier behaviour in a radwaste disposal, *Appl. Geochem.* 20(12): 2252–2282.

Pifer, A. D., Miskin, d. R., Cousins, S. L. & Fairey, J. L. (2011). Coupling asymmetric flow-field flow fractionation and fluorescence parallel factor analysis reveals stratification of dissolved organic matter in a drinking water reservoir, *J. Chromatogr. A* 1218(27): 4167–4178.

Pokrovsky, O. S. & Schott, J. (2002). Iron colloids/organic matter associated transport of major and trace elements in small boreal rivers and their estuaries (NW Russia), *Chem. Geol.* 190: 141–179.

Popic, J. M., Salbu, B., Strand, T. & Skipperud, L. (2011). Assessment of radionuclide and metal contamination in a thorium rich area in Norway, *J. Environ. Monit.* 13: 1730–1738.

Porcelli, D., Andersson, P. S., Wasserburg, G. J., Ingri, J. & Baskaran, M. (1997). The importance of colloids and mires for the transport of uranium isotopes through the Kalix River watershed and Baltic Sea, *Geochim. Cosmochim. Acta* 61(19): 4095–4113.

Pourret, O., Dia, A., Davranche, M., Gruau, G., Hénin, O. & Angée, M. (2007). Organo-colloidal control on major- and trace-element partitioning in shallow groundwaters: Confronting ultrafiltration and modelling, *Appl. Geochem.* 22(8): 1568–1582.

Pourret, O., Gruau, G., Dia, A., Davranche, M. & Molénat, J. (2010). Colloidal control on the distribution of rare earth elements in shallow groundwaters, *Aquatic Geochem.* 16(1): 31–59.

Pshinko, G. N. (2009). Impact of humic matter on sorption of radionuclides by montmorrilonite, *J. Water Chem. Technol.* 31: 163–171.

Pshinko, G. N., Timoshenko, T. G. & Bogolepov, A. A. (2009). Effect of fulvic acids on Th(IV) sorption on montmorillonite, *Radiochemistry* 51: 91–95.

Qureshi, R. N. & Kok, W. T. (2011). Application of flow field-flow fractionation for the characterization of macromolecules of biological interest: a review, *Anal. Bioanal. Chem.* 399(4): 1401–1411.

Rabung, T., Pierret, M. C., Bauer, A., Geckeis, H., Bradbury, M. H. & Baeyens, B. (2005). Sorption of Eu(III)/Cm(III) on Ca-montmorillonite and Na-illite. Part 1: Batch sorption and time-resolved laser fluorescence spectroscopy experiments, *Geochim. Cosmochim. Acta* 69(23): 5393–5402.

Rachkova, N. G., Shuktomova, I. I. & Taskaev, A. I. (2010). The state of natural radionuclides of uranium, radium, and thorium in soils, *Eurasian Soil Sci.* 43: 651–658.

Ranville, J. F., Hendry, M. J., Reszat, T. N., Xie, Q. & Honeyman, B. D. (2007). Quantifying uranium complexation by groundwater dissolved organic carbon using asymmetrical flow field-flow fractionation, *J. Contam. Hydrol.* 91(3-4): 233–246.

Sabodina, M. N., Kalmykov, S. N., Artem'eva, K. A., Zakharova, E. V. & Sapozhnikov, Y. A. (2006). Behavior of Cs, Np(V), Pu(IV), and U(VI) in pore water of bentonite, *Radiochemistry* 48: 488–492.

Salbu, B. (2007). Speciation of radionuclides – analytical challenges within environmental impact and risk assessments, *J. Environ. Radioact.* 96(1-3): 47–53.

Salbu, B. (2009). Fractionation of radionuclide species in the environment, *J. Environ. Radioact.* 100(4): 283–289.

Salbu, B., Lind, O. C. & Skipperud, L. (2004). Radionuclide speciation and its relevance in environmental impact assessments, *J. Environ. Radioact.* 74(1-3): 233–242.

Salbu, B. & Skipperud, L. (2009). Speciation of radionuclides in the environment, *J. Environ. Radioact.* 100(4): 281–282.

Santos, A. & Barros, P. H. L. (2010). Multiple particle retention mechanisms during filtration in porous media, *Environ. Sci. Technol.* 44(7): 2515–2521.

Schelero, N. & von Klitzing, R. (2011). Correlation between specific ion adsorption at the air/water interface and long-range interactions in colloidal systems, *Soft Matter* 7: 2936–2942.

Schindler, M., Fayek, M. & Hawthorne, F. C. (2010). Uranium-rich opal from the Nopal I uranium deposit, Peña Blanca, Mexico: Evidence for the uptake and retardation of radionuclides, *Geochim. Cosmochim. Acta* 74(1): 187–202.

Schmeide, K. & Bernhard, G. (2010). Sorption of Np(V) and Np(IV) onto kaolinite: Effects of pH, ionic strength, carbonate and humic acid, *Appl. Geochem.* 25(8): 1238–1247.

Scholten, J. C., Fietzke, J., Mangini, A., Stoffers, P., Rixen, T., Gaye-Haake, B., Blanz, T., Ramaswamy, V., Sirocko, F., Schulz, H. & Ittekkot, V. (2005). Radionuclide fluxes in the Arabian Sea: the role of particle composition, *Earth Planet. Sci. Lett.* 230(3–4): 319–337.

Seiler, R. L., Stillings, L. L., Cutler, N., Salonen, L. & Outola, I. (2011). Biogeochemical factors affecting the presence of [210]Po in groundwater, *Appl. Geochem.* 26(4): 526–539.

Seliman, A. F., Borai, E. H., Lasheen, Y. F., Abo-Aly, M. M., DeVol, T. A. & Powell, B. A. (2010). Mobility of radionuclides in soil/groundwater system: Comparing the influence of EDTA and four of its degradation products, *Environ. Pollut.* 158(10): 3077–3084.

Semizhon, T., Röllin, S., Spasova, Y. & Klemt, E. (2010). Transport and distribution of artificial gamma-emitting radionuclides in the River Yenisei and its sediment, *J. Environ. Radioact.* 101(5): 385–402.

Severino, G., Cvetkovic, V. & Coppola, A. (2007). Spatial moments for colloid-enhanced radionuclide transport in heterogeneous aquifers, *Adv. Water Resour.* 30(1): 101–112.

Singer, D. M., Farges, F. & Brown Jr, G. E. (2009). Biogenic nanoparticulate UO_2: Synthesis, characterization, and factors affecting surface reactivity, *Geochim. Cosmochim. Acta* 73(12): 3593–3611.

Singer, D. M., Maher, K. & Brown Jr, G. (2009). Uranyl-chlorite sorption/desorption: Evaluation of different U(VI) sequestration processes, *Geochim. Cosmochim. Acta* 73(20): 5989–6007.

Singh, B. K., Jain, A., Kumar, S., Tomar, B. S., Tomar, R., Manchanda, V. K. & Ramanathan, S. (2009). Role of magnetite and humic acid in radionuclide migration in the environment, *J. Contam. Hydrol.* 106(3-4): 144–149.

Singhal, R. K., Karpe, R., Muthe, K. P. & Reddy, A. V. R. (2009). Plutonium-239+240 selectivity for pseudo-colloids of iron in subsurface aquatic environment having elevated level of dissolved organic carbon, *J. Radioanal. Nucl. Chem.* 280: 141–148.

Solovitch-Vella, N., Garnier, J.-M. & Ciffroy, P. (2006). Influence of the colloid type on the transfer of ^{60}Co and [85]Sr in silica sand column under varying physicochemical conditions, *Chemosphere* 65(2): 324–331.

Stepanets, O. V., Ligaev, A. N., Borisov, A. P., Travkina, A. V., Shkinev, V. M., Danilova, T. V., Miroshnikov, A. Y. & Migunov, V. I. (2009). Geoecological investigations of the Ob-Irtysh River Basin in the Khanty-Mansi Autonomous region: Yugra in 2006–2007, *Geochem. Int.* 47: 657–671.

Stolpe, B., Hassellöv, M., Andersson, K. & Turner, D. R. (2005). High resolution ICPMS as an on-line detector for flow field-flow fractionation; multi-element determination of colloidal size distributions in a natural water sample, *Anal. Chim. Acta* 535(1-2): 109–121.

Stumm, W. & Morgan, J. J. (1996). *Aquatic Chemistry: Chemical Equilibria and Rates in Natural Waters*, 3rd edn, John Wiley & Sons, Ltd.

Tang, X.-Y. & Weisbrod, N. (2009). Colloid-facilitated transport of lead in natural discrete fractures, *Environ. Pollut.* 157(8-9): 2266–2274.

Tang, X.-Y. & Weisbrod, N. (2010). Dissolved and colloidal transport of cesium in natural discrete fractures, *J. Environ. Qual.* 39(3): 1066–1076.

Templeton, D. M., Ariese, F., Cornelis, R., Danielsson, L.-G., Muntau, H., Leeuwen, H. P. & Lobiński, R. (2000). Guidelines for terms related to chemical speciation and fractionation of elements. Definitions, structural aspects, and methodological approaches (IUPAC Recommendations 2000), *Pure Appl. Chem.* 72(8): 1453–1470.

Tertre, E., Berger, G., Castet, S., Loubet, M. & Giffaut, E. (2005). Experimental sorption of Ni^{2+}, Cs^+ and Ln^{3+} onto a montmorillonite up to 150°C, *Geochim. Cosmochim. Acta* 69(21): 4937–4948.

Utsunomiya, S., Kersting, A. B. & Ewing, R. C. (2009). Groundwater nanoparticles in the far-field at the Nevada test site: Mechanism for radionuclide transport, *Environ. Sci. Technol.* 43(5): 1293–1298.

van der Veeken, P. L. R., Pinheiro, J. P. & van Leeuwen, H. P. (2008). Metal speciation by DGT/DET in colloidal complex systems, *Environ. Sci. Technol.* 42(23): 8835–8840.

Vandenhove, H., Antunes, K., Wannijn, J., Duquène, L. & Van Hees, M. (2007). Method of diffusive gradients in thin films (DGT) compared with other soil testing methods to predict uranium phytoavailability, *Sci. Total Environ.* 373(2-3): 542–555.

Vilks, P., Miller, N. & Vorauer, A. (2008). Laboratory bentonite colloid migration experiments to support the Äspö colloid project, *Phys. Chem. Earth, Pts. A/B/C* 33(14-16): 1035–1041.

Vrana, B., Allan, I. J., Greenwood, R., Mills, G. A., Dominiak, E., Svensson, K., Knutsson, J. & Morrison, G. (2005). Passive sampling techniques for monitoring pollutants in water, *Trends Anal. Chem.* 24(10): 845–868.

Walther, C., Rothe, J., Brendebach, b., Fuss, M., Altmaier, M., Marquardt, C. M., Büchner, S., Cho, H.-R., Yun, J.-I. & Seibert, A. (2009). New insights in the formation processes of Pu(IV) colloids, *Radiochim. Acta* 97: 199–207.

Wang, X., Tan, X., Ning, Q. & Chen, Q. (2005). Simulation of radionuclides ^{99}Tc and ^{243}Am migration in compacted bentonite, *Appl. Radiat. Isot.* 62(5): 759–764.

Warnken, K. W., Zhang, H. & Davison, W. (2004). Analysis of polyacrylamide gels for trace metals using diffusive gradients in thin films and laser ablation inductively coupled plasma mass spectrometry, *Anal. Chem.* 76(20): 6077–6084.

Wersin, P., Stroes-Gascoyne, S., Pearson, F. J., Tournassat, C., Leupin, O. X. & Schwyn, B. (2011). Biogeochemical processes in a clay formation *in situ* experiment: Part G - Key interpretations and conclusions. implications for repository safety, *Appl. Geochem.* 26(6): 1023 – 1034.

Wieland, E., Tits, J. & Bradbury, M. H. (2004). The potential effect of cementitious colloids on radionuclide mobilisation in a repository for radioactive waste, *Appl. Geochem.* 19: 119–135.

Wigginton, N. S., Haus, K. L. & Hochella Jr, M. F. (2007). Aquatic environmental nanoparticles, *J. Environ. Monit.* 9(12): 1306–1316.

Wilkins, M. J., Livens, F. R., Vaughan, D. J. & Lloyd, J. R. (2006). The impact of Fe(III)-reducing bacteria on uranium mobility, *Biogeochemistry* 78: 125–150.

Wilkins, M. J., Livens, F. R., Vaughan, D. J., Lloyd, J. R., Beadle, I. & Small, J. S. (2010). Fe(III) reduction in the subsurface at a low-level radioactive waste disposal site, *Geomicrobiol. J.* 27(3): 231–239.

Williams, S. K. R., Runyon, J. R. & Ashames, A. A. (2011). Field-flow fractionation: Addressing the nano challenge, *Anal. Chem.* 83(3): 634–642.

Wolthoorn, A., Temminghoff, E. J. M., Weng, L. & van Riemsdijk, W. (2004). Colloid formation in groundwater: effect of phosphate, manganese, silicate and dissolved organic matter on the dynamic heterogeneous oxidation of ferrous iron, *Appl. Geochem.* 19(4): 611–622.

Wu, Z., He, M. & Lin, C. (2011). In situ measurements of concentrations of Cd, Co, Fe and Mn in estuarine porewater using DGT, *Environ. Poll.* 159(5): 1123–1128.

Xu, C., Zhang, S., Ho, Y.-F., Miller, E. J., Roberts, K. A., Li, H.-P., Schwehr, K. A., Otosaka, S., Kaplan, D. I., Brinkmeyer, R., Yeager, C. M. & Santschi, P. H. (2011). Is soil natural organic matter a sink or source for mobile radioiodine (^{129}I) at the Savannah River Site?, *Geochim. Cosmochim. Acta* 75(19): 5716–5735.

Yamaguchi, T., Nakayama, S., Vandergraaf, T. T., Drew, D. J. & Vilk (2008). Radionuclide and colloid migration experiments in quarried block of granite unde in-situ coditions at a depth of 240 m, *J. Power Energy Syst.* 2: 186–197.

Yeager, K. M., Santschi, P. H., Phillips, J. D. & Herbert, B. E. (2005). Suspended sediment sources and tributary effects in the lower reaches of a coastal plain stream as indicated by radionuclides, Loco Bayou, Texas, *Environ. Geol.* 47: 382–395.

Yoshida, T. & Suzuki, M. (2006). Migration of strontium and europium in quartz sand column in the presence of humic acid: Effect of ionic strength, *J. Radioanal. Nucl. Chem.* 270: 363–368.

Zhang, H. & Davison, W. (2000). Direct in situ measurements of labile inorganic and organically bound metal species in synthetic solutions and natural waters using diffusive gradients in thin films, *Anal. Chem.* 72(18): 4447–4457.

Zhang, H. & Davison, W. (2001). *In situ* speciation measurements. Using diffusive gradients in thin films (DGT) to determine inorganically and organically complexed metals, *Pure Appl. Chem.* 73(1): 9–15.

7

An Assessment
of the Impact of Advanced
Nuclear Fuel Cycles on Geological Disposal

Jan Marivoet and Eef Weetjens
Belgian Nuclear Research Centre (SCK•CEN)
Belgium

1. Introduction

After a strong growth in the 1970s and 1980s, the deployment of nuclear energy has stagnated. Only a few nuclear power plants became operational during the last 20 years. More recently, rising oil prices, increasing energy demands and the need for carbon-free energy to limit global warming have resulted in a revival of interests in nuclear energy. New nuclear power plants are now under construction in, e.g., Finland, France, Japan, Korea, the Russian Federation, China and India. Especially in these last 2 countries a considerable increase of the contribution of nuclear energy to the electricity generation is expected (Nuclear Energy Agency [NEA], 2008). The nuclear renaissance made that advanced nuclear fuel cycles are being studied worldwide. They aim at making more efficient use of the available resources, reducing the risk of proliferation of nuclear weapons, and facilitating the management of the radioactive waste. However, the Fukushima accident in 2011 strongly tempered the revival of nuclear energy in several countries.

During the last 20 years the introduction of partitioning and transmutation (P&T) techniques in future nuclear fuel cycles has often been presented as a promising option that might strongly facilitate the long-term management of high-level radioactive waste, because transmutation of the actinides, which are present in spent nuclear fuel, in fast neutron reactors can strongly reduce the long-term radiotoxicity of the radioactive waste. On the other hand, the relevance of this reduction of radiotoxicity for radioactive waste management is questioned by the geological disposal community.

The impact of advanced fuel cycles on radioactive waste management and geological disposal has been studied in some national and international projects between 2003 and 2010. Highly relevant projects were the Expert Group on the Impact of P&T on Radioactive Waste Management of the Nuclear Energy Agency (NEA, 2006) and the Red-Impact project of the European Commission (von Lenza et al., 2008; Marivoet et al., 2009). In the following sections the impact of advanced fuel cycles on various aspects of radioactive waste management and, more specifically, of geological disposal in a clay formation will be investigated. The presented results were mainly obtained from SCK•CEN's participation in the Red-Impact project (Marivoet et al., 2007; Weetjens & Marivoet, 2008).

2. Considered fuel cycles and repository system

Making an assessment of the impact of advanced fuel cycles on radioactive waste management and geological disposal requires that a number of representative fuel cycle scenarios are identified. As a basis for comparison, various indicators are presented in the following sections by taking disposal in Boom Clay as example to develop the case-study. Results obtained in international projects (NEA, 2006; Marivoet et al., 2009) show that most conclusions that can be drawn from the evaluations of disposal in Boom Clay are largely valid for disposal in other clay formations and to a large extent for disposal in hard rock formations.

2.1 Considered fuel cycles

On the basis of an NEA report on advanced fuel cycles (NEA, 2002), 5 fuel cycle scenarios (3 industrial scenarios, which are based on presently available technology, and 2 innovative scenarios) were selected for detailed analyses in the Red-Impact project:

- fuel cycle A1: the reference fuel cycle, which is the "once through" cycle based on pressurised water reactor (PWR) plants with uranium oxide (UOX) fuel;
- fuel cycle A2: fuel cycle based on PWR plants with uranium oxide fuel, the generated Pu is recycled once as mixed oxide (MOX) fuel;
- fuel cycle A3: fuel cycle based on a sodium cooled fast neutron reactors with MOX fuels, in which Pu is multi-recycled;
- fuel cycle B1: fuel cycle based on a sodium cooled fast neutron reactors with MOX fuels, in which all the actinides are recycled;
- fuel cycle B2: fuel cycle based on PWR plants with uranium oxide fuel, the generated Pu is recycled once as MOX fuel, all the minor actinides and the Pu in the spent MOX fuel are recycled in a fast neutron accelerator driven system (ADS).

Results obtained for a variant of fuel cycle B1 (called B1V), in which it is assumed that the heat generating Cs and Sr are separated from the high-level waste (HLW), are also presented.

For each selected fuel cycle mass flow schemes have been prepared and the corresponding neutronic calculations have been made (von Lensa et al., 2008). The estimated waste inventories strongly depend on the assumed separation efficiency of the considered partitioning processes and the applied waste conditioning methods. It was assumed that in future reprocessing plants the reprocessing losses will be limited to 0.1% for all the actinides. For the volatile fission and activation products it was assumed that 1% of the ^{129}I and ^{36}Cl and 10% of the ^{14}C remain in the high-level waste. Specific and optimised waste matrices for conditioning the high-level waste that will arise from advanced fuel cycles have not yet been developed. Therefore, it was assumed that the conditioning methods for the waste from advanced fuel cycles will be very similar to the ones applied today in existing reprocessing facilities.

The volumes of high-level (HLW) and intermediate-level waste (ILW) that are generated in the considered fuel cycles have been estimated in the Red-Impact project (Cunado, 2007) as follows. For the 2 fuel cycles that are applied today (A1 and A2) realistic estimates of the generated waste volumes could be made on the basis of data provided by the participating waste agencies. For the future fuel cycles (A3, B1 and B2) the volumes of the HLW were estimated on the basis of the mass of the generated fission products and assumed reprocessing losses, and by assuming that the waste will be conditioned as vitrified HLW with the same waste loading as applied today for cycle A2. The volumes of ILW were estimated on the basis of the information available on the cladding and structural components of the fuel assemblies, whereas for the technological waste from reprocessing it was assumed that future reprocessing plants will generate similar waste volumes as the present ones.

The amount of natural uranium that is needed in each of the considered fuel cycles to produce 1 TWh(e) electricity is given in Table 1 (Marivoet & Weetjens, 2008). This indicator illustrates how the introduction of advanced fuel cycles increases the efficiency of the use of uranium. The recycling of Pu in fuel cycle A2 results in a small increase, of about 10%, of the efficiency of the use of uranium for electricity production. Recycling Pu and introducing an ADS in a nuclear reactor park consisting of PWRs (fuel cycle B2) results in an increase of the generated electricity with about 25%. However, replacing PWRs by fast neutron reactors results in a drastic increase of the amount of electricity produced per kg uranium, because natural or depleted uranium can be used instead of enriched uranium for the production of the reactor fuel. In principle, fully recycling all the actinides in a fast reactor (fuel cycle B1) can result in an increase of the efficiency of the use of natural uranium with a factor 100 or more.

Fuel cycle		A1	A2	A3	B1	B2
Consumption of natural uranium	kg/TWh(e)	20723	18448	986	106	15766

Table 1. Amount of natural uranium needed to produce 1 TWh(e) electricity (Marivoet & Weetjens, 2008)

2.2 Considered repository system

The repository considered for the impact evaluations is the SAFIR 2 repository concept that has been developed by the Belgian radioactive waste management organisation ONDRAF/NIRAS (2002). This reference repository is assumed to be excavated in the Boom Clay formation at the Mol site in NE Belgium. At the considered site the host formation is about 100 m thick, of which an 80 m thick central zone consists of very homogeneous clay (in terms of its flow and transport parameters).

The considered repository will have a central access facility consisting of at least two vertical transport shafts and two horizontal transport galleries. The disposal galleries will be excavated perpendicular upon the transport galleries. The HLW canisters will be placed one after the other or with some spacing between two canisters to respect temperature limitations, along the axis of the disposal galleries. A clayey buffer will be placed between the HLW canisters and the gallery walls. Because the Boom Clay is a plastic clay, a concrete lining is required to avoid convergence of the gallery walls during the operational phase of the repository. Figure 1 shows a scheme of a disposal gallery configuration for uranium oxide spent fuel.

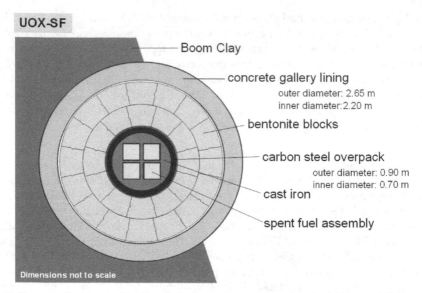

Fig. 1. Gallery configuration for disposal of uranium oxide spent fuel (4 assemblies per container)

3. Assessment of the impact of advanced fuel cycles

3.1 Impact on waste volumes

The total volumes of the waste canisters (these are the primary waste containers) containing the generated HLW and ILW are given in Table 2 (Cunado, 2007). These results show that the volume of the HLW generated in fuel cycle B1 is 5 times smaller than the volume of the spent fuel from fuel cycle A1. However, when the volume of the ILW is also considered, the volume reduction is limited to a factor 2.

Fuel cycle		A1	A2	A3	B1	B2
HLW volume	m³/TWh(e)	2.216	0.806	0.420	0.397	0.464
ILW volume	m³/TWh(e)	-	0.387	0.825	0.825	0.518
Total volume	m³/TWh(e)	2.216	1.193	1.245	1.223	0.982
Relative volume	-	-	1.000	0.539	0.562	0.552	0.443

Table 2. Volumes of the generated HLW and ILW (packed in canisters) estimated for the considered fuel cycles

The actual volume of HLW is not considered a meaningful indicator for geological disposal. If waste volumes were important, it would be relatively simple to reduce the volume of spent fuel by removing the void between the fuel pins by dismantling the assemblies. However, the dimensions of a geological repository for HLW disposal are mainly imposed by the thermal output of the waste and not by its volume. Therefore, the thermal output and the dimensions of the repository are discussed in the following section.

3.2 Impact on repository dimensions

Figure 2 gives the evolution of the thermal output of the HLW generated in the considered fuel cycles. After a 50 years cooling time the thermal output of the vitrified HLW from cycle A2 is only 57% of the one of spent fuel from cycle A1; however, if the MOX spent fuel from cycle A2 has to be disposed of as well, the reduction of the thermal output is only 3%. In the case of advanced fuel cycle B1 the thermal output of the HLW is reduced with a factor 3 in comparison with the reference cycle A1. It is also interesting to note that the thermal output of the HLW from fuel cycle B1 is 3 times lower after a 100 years cooling time than after a 50 years cooling time; over the same time span the thermal output of the spent fuel from cycle A1 is only reduced with a factor 2.

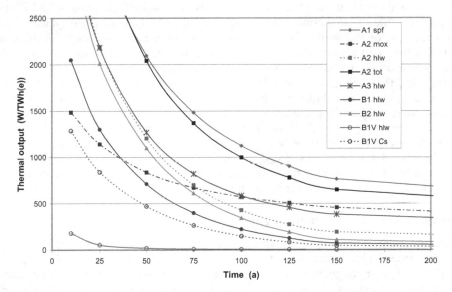

Fig. 2. Evolution of the thermal output of the high-level waste resulting from the considered fuel cycles (cycle B1V refers to cycle B1 with removal of Cs and Sr from the HLW) (Marivoet & Weetjens, 2009)

Removal of the heat generating Cs and Sr from the HLW results in a strong reduction of the thermal output of that waste; e.g., in the case of fuel cycle B1 the thermal output of the HLW is reduced with a factor 40. However, in this case the strongly heat generating Cs and Sr waste has to be stored during at least a century in a cooling facility to have a significant reduction of the thermal output, and eventually the Cs-waste, which contains the very long-lived ^{135}Cs ($T_{1/2}$ = 2.3 Ma), has to be disposed of in a geological repository.

The minimum lengths of the galleries for disposal of HLW are derived from heat dissipation calculations by ensuring that the temperature limitations are respected; for the considered disposal concept the main temperature limitation is that the temperature at the interface between the gallery liner and the host clay formation should remain below 100 °C. An overview of the estimated lengths of the HLW disposal galleries for a 50 years cooling time is given in Table 3.

Fuel cycle		A1	A2	A3	B1	B1V	B2
Allowable thermal loading (50 a)	W/m	353	332-376	365	379	-	379
HLW gallery length	m/TWh(e)	5.92	5.74	3.48	1.88	0.44	2.89
Relative HLW gallery length	-		1.00	0.97	0.59	0.32	0.074	0.49

Table 3. Allowable thermal loading per metre gallery and estimated minimum lengths of the HLW disposal galleries for a 50 years cooling time (cycle B1V refers to a variant of cycle B1 with removal of Cs and Sr from the HLW; a 100 years cooling time is assumed for the Cs-waste) (Marivoet & Weetjens, 2009)

For HLW containing only traces of actinides, which is the case for fuel cycles B1 and B2, the maximum allowable linear thermal loading at disposal time of the disposal gallery is 379 W/m; in the case of the MOX spent fuel from cycle A2, which contains a considerable fraction of actinides, this maximum thermal loading is reduced to 332 W/m. Because of the relatively small difference (about 12%) between the 2 extreme values of the allowable thermal loading, the minimum lengths of the disposal galleries are about proportional to the thermal output of the waste at disposal time. Multi-recycling of the Pu in fast reactors (cycle A3) results in a reduction of the needed length of the disposal galleries of about 40%, a PWR reactor park combined with an ADS for burning the minor actinides and the multi-recycled Pu (cycle B2) gives a reduction of about 50%, and full recycling of all actinides in fast reactors gives a reduction of about 68%. By assuming that the Cs-waste is cooled during 100 years, the removal of Cs and Sr from the HLW gives a reduction of the needed gallery length of more than 90% in comparison with the once through cycle A1.

3.3 Evolution of radiotoxicity in the waste

Radiotoxicity is a measure of how nocuous a radionuclide is to health. The type and energy of rays, absorption in the organism, residence time in the body, etc. influence the degree of radiotoxicity of a radionuclide. In the context of radioactive waste management radiotoxicity is defined as the sum over all radionuclides present in the waste of the products of the activity of a radionuclide with its dose coefficient for ingestion; values of the latter are published by the International Commission on Radiological Protection (ICRP, 1996).

The introduction of P&T techniques in a fuel cycle, i.e. recycling of all actinides in a fast neutron spectrum facility (fast reactor or ADS), results in a drastic reduction of the radiotoxicity in the high-level waste (see Figure 3).

After about 30 years cooling time the minor actinides, mainly Pu, Am and Cm isotopes, are the main contributors to the radiotoxicity. In the case of the reference fuel cycle A1 it takes more than 200 000 years before the radiotoxicity in the spent fuel has dropped below the level present in the natural uranium that was needed to produce that fuel. For fuel cycle A2 only a small reduction of the radiotoxicity can be observed because all the minor actinides and the Pu of the MOX spent fuel are present in the disposed waste. Recycling of all the generated Pu (cycle A3) makes that the radiotoxicity of the natural uranium is reached after about 20 000 years, and when all the actinides are recycled (cycles B1 and B2) it is reached already after about 400 years.

The strong reduction of the time after which radiotoxicity reaches the level of natural uranium is one of the main arguments used to justify the introduction of P&T in future fuel cycles. However, as we will see in the following section, the actinides are very efficiently confined in the geological disposal system and their releases into the biosphere are about negligible, because most actinides are poorly soluble in the chemical conditions prevailing in a geological repository in clay and, furthermore, all actinides are strongly sorbed on the clay minerals (or immobile organic matter) of the host formation. As a consequence, there is no relationship between radiotoxicity and the radiological consequences of a geological repository.

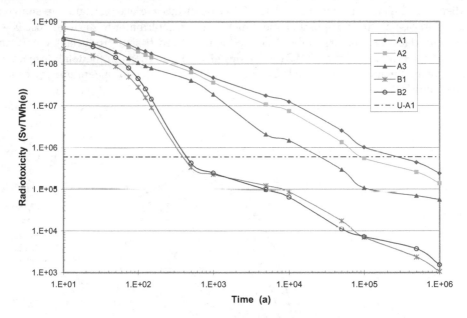

Fig. 3. Evolution of the radiotoxicity present in the high-level waste arising from the 5 considered fuel cycles (the reference line U-A1 gives the radiotoxicity present in the natural uranium needed to generate 1 TWh(e) in cycle A1) (Marivoet & Weetjens, 2009)

3.4 Impact on long-term dose in the case of the reference scenario

The main indicator to evaluate the safety of a geological repository for the disposal of radioactive waste is the effective annual dose to a representative person of the potentially most exposed group in the case of the reference (i.e. expected evolution of the repository system) scenario. The International Commission on Radiological Protection (ICRP, 2000) recommends to use a dose constraint of 0.3 mSv/a, or lower, as reference value for the effective dose. Most national radiological protection authorities are now imposing a dose constraint in the range 0.1 – 0.3 mSv/a.

The reference scenario assumes that the barriers of the disposal system will function as expected. For the considered repository system it was assumed that the overpack will have a lifetime of 2000 years, and that the waste matrices of the vitrified HLW and spent fuels will dissolve in a 100 000 and 200 000 years period respectively. The solubility limits in the near field are taken into account. The main barrier of the disposal system is the natural barrier provided by the host clay formation; because water flow in the clay is negligible, the radionuclide transport through the clay layer is essentially diffusive; furthermore, many radionuclides are strongly sorbed on the clay minerals. After migration through the clay layer, only a small fraction of the disposed radionuclides reach the aquifers overlying the host formation and eventually the biosphere via a small river draining the contaminated part of the aquifer.

The total dose (normalised per produced electricity) and its main contributors calculated for the disposal of spent fuel from fuel cycle A1 in case of the reference scenario are shown in Figure 4. In case of direct disposal of spent fuel the main contributor to the total dose is ^{129}I. Other important contributors to the dose are the long-lived fission products ^{79}Se, ^{99}Tc and ^{126}Sn and the activation product ^{36}Cl. Actinides give contributions to the total dose after 2 million years; the maximum actinide dose is more than 2 orders of magnitude lower than the maximum ^{129}I dose.

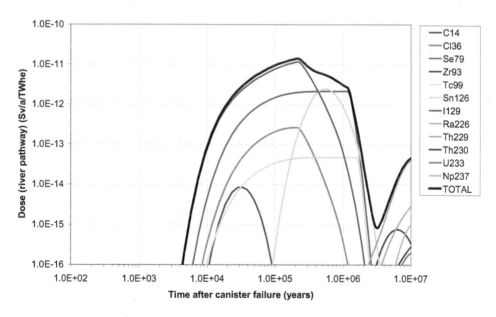

Fig. 4. Evolution of the total effective dose via the river pathway due to the disposal of spent fuel arising from fuel cycle A1 in the case of the reference scenario

Figure 5 gives the doses calculated for the disposal of the high-level and intermediate-level waste (ILW) in case of the reference scenario for the 5 considered fuel cycles. The calculated total doses (estimated by multiplying the normalised doses with the electricity generated within the existing national nuclear programme; the present Belgian nuclear programme is expected to generate about 1700 TWh(e)) are a few orders of magnitude below the dose constraint.

The impact of the transmutation of the actinides in the advanced fuel cycles on the maximum doses is negligible, because the maximum dose is essentially due to mobile long-lived fission and activation products. The amount of generated fission products is about proportional to the heat produced by nuclear fission in the reactors. The difference in the calculated maximum doses is mainly explained by the amount of [129]I that is present in the disposed waste. In the case of spent fuel disposal (cycle A1, and partially in the case of cycle A2 in which the MOX spent fuel is assumed to be directly disposed of), all the iodine is going into the repository, whereas in the case of reprocessing only a small fraction of the iodine, which is estimated to be about 2% (1% in the HLW and 1% in the ILW), remains in the waste; the bulk of the iodine escapes as gas from the HLW stream in the dissolver at the reprocessing plant. On the very long-term, i.e. after a few million years for disposal in clay, calculated doses are lower in the case of advanced fuel cycles, because smaller amounts of actinides are present in the HLW arising from these fuel cycles.

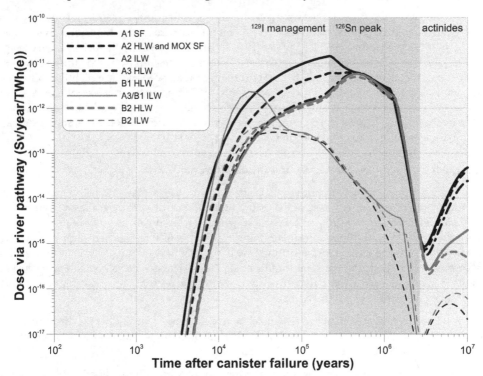

Fig. 5. Evolution of the total effective dose due to the disposal of the high-level and intermediate-level waste arising from the 5 considered fuel cycles (Marivoet & Weetjens, 2008)

Figure 5 also gives the doses due to the disposal of the ILW arising from the considered fuel cycles. For fuel cycle A2 the maximum ILW dose is more than one order of magnitude lower than the maximum HLW dose. However, the ILW dose due to cycles A3 and B1, which are based on fast reactors, shows a relatively high initial dose. This dose is due to the considerable amounts of ^{14}C which are calculated to be generated in fast reactors.

3.5 Impact on long-term dose in other cases and altered evolution scenarios

In the safety assessments made for the SAFIR 2 report and for the upcoming Belgian safety case a number of altered evolution scenarios have been analysed as well. Also for those scenarios, the total dose is mainly due to mobile fission and activation products. However, actinides can play a more important role in safety assessments of disposal in other host formations, such as hard rock (granite) and volcanic rocks (tuff or basalt).

In the case of Yucca Mountain in the United States (Department of Energy [DoE], 2008), the spent fuel is assumed to be disposed above the groundwater table in a mountain consisting of welded tuff located in an arid area. A very low amount of water is flowing through the repository. The oxidising chemical conditions in the disposal galleries result in a relatively high solubility of some actinides. As a consequence the total dose calculated for Yucca Mountain is due to mobile fission and activation products and to actinides.

Disposal in hard rock formations is studied in, e.g., Sweden and Finland and safety cases have been submitted to the radiological protection authorities by the respective waste management organisations (SKB, 2006; POSIVA, 2010). In these cases the use of a very long-lived copper canister makes that no releases of radionuclides are expected during the first hundreds of thousands of years. In the case of the reference scenario in which the failure of some canisters is assumed, the dose is mainly due to ^{129}I and to a lesser extent to ^{135}Cs and ^{14}C. However, for several variant scenarios, such as penetration of oxidising glacial melt water in the near field or a degradation of the bentonite buffer, ^{226}Ra, ^{231}Pa and ^{239}Pu can become the dominant dose contributors.

3.6 Impact on the dose in the case of a human intrusion

A few very low probability disruptive scenarios can have the potential to bring man directly in contact with the disposed waste. One of the most drastic disruptive scenarios that might be considered is a human intrusion scenario in which it is assumed that a borehole is drilled across the repository and that cores containing fragments of the disposed high-level waste are taken from that borehole and examined by a geotechnical worker, who is not aware of the presence of radioactive material (Smith et al., 1987). The dose to a geotechnical worker and its contributors calculated for an intrusion into the repository containing spent fuel from fuel cycle A1 are shown in Figure 6.

Up to 100 000 years minor actinides, successively ^{241}Am, ^{240}Pu and ^{239}Pu, are the main contributors to the dose to a geotechnical worker; at the long term, i.e. after 100 000 years, the main contributors to the dose are ^{222}Rn and ^{229}Th.

The doses to a geotechnical worker calculated for an intrusion into the repository containing the HLW arising from the considered fuel cycles are shown in Figure 7. This figure also gives 3 reference lines, which correspond to the ICRP intervention levels of 10 and 100 mSv and to the dose calculated for a geotechnical worker who is handling and analysing cores that are assumed to be taken from the uranium mine at Cigar Lake (Canada).

Figure 7 shows that only for 3 HLW types, arising from the advanced fuel cycles B1 and B2, the dose to a geotechnical worker drops within 1 million years under the dose associated with an intrusion into a Cigar Lake uranium ore body. For the other high-level waste and spent fuel types the doses to a geotechnical worker remain (much) higher than the dose associated with an intrusion into the Cigar Lake ore body. Consideration of the ICRP intervention levels of 10 and 100 mSv (ICRP, 2000) can be used to contextualise the dose values calculated for the various spent fuel and vitrified HLW types. Applying a simple criterion of ranking when the calculated dose for each HLW type is bounded by the ICRP intervention levels allows a "ranking" of the high-level waste types.

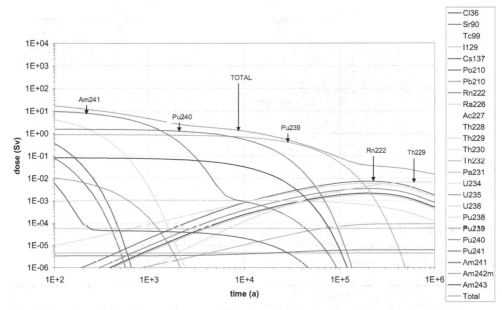

Fig. 6. Evolution of the dose to a geotechnical worker and its contributors calculated for an intrusion into a repository containing spent fuel from fuel cycle A1

Table 4 gives the estimated "required isolation times" for the main spent fuel or vitrified HLW types arising from each of the 5 considered fuel cycle scenarios. In order to verify whether the radiotoxicity present in the disposed waste can be used as an alternative indicator, Table 4 also gives the time after which the radiotoxicity in the disposed waste drops under the radiotoxicity in fresh uranium required for the fuel production in fuel cycle A1 (see Fig. 7). Comparison of the times obtained in Table 4 from the 100 mSv intervention level with the times derived from the radiotoxicity shows that these times are often of the same order of magnitude.

In the case of fuel cycles generating different HLW types, one or more waste types can occur in which a lot of radioactivity is concentrated, but of which only a very small number of waste packages are generated; this is the case for the MOX spent fuel in fuel cycle A2 and the vitrified HLW arising from the pyro-reprocessing of ADS spent fuel in fuel cycle B2. The results given in Table 4 show that the "required isolation times" obtained for the 100 mSv intervention level for the waste types of which the largest quantities are generated correspond much better with the times estimated on the basis of the radiotoxicity arising from the considered fuel cycle.

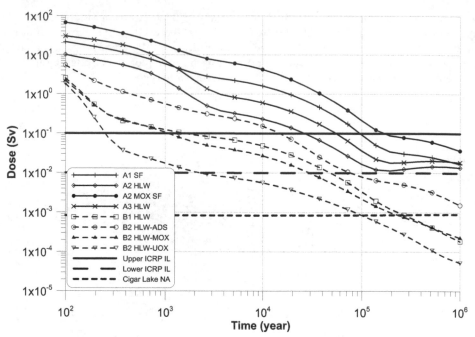

(IL: intervention level; NA: natural analogue)

Fig. 7. Evolution of the dose to a geotechnical worker for the 5 considered fuel cycles
(Marivoet & Weetjens, 2009)

Fuel cycle	Comparator / Waste type	ICRP 10 mSv intervention level	ICRP 100 mSv intervention level	Radiotoxicity
Fuel cycle B1	HLW	~ 40 000 a	~ 1000 a	~ 300 a
Fuel cycle B2	HLW-UOX	~ 2000 a	~ 250 a	~ 300 a
	HLW-ADS	~ 70 000 a	~ 13 000 a	
Fuel cycle A3	HLW	> 1 Ma	~ 70 000 a	~ 24 000 a
Fuel cycle A2	HLW	~ 100 000 a	~ 20 000 a	~ 90 000 a
	MOX SF	> 1 Ma	~ 200 000 a	
Fuel cycle A1	UOX SF	> 1 Ma	~ 100 000 a	~ 200 000 a

Table 4. Ranking of high-level waste and spent fuel types, on basis of calculated dose from
geotechnical worker scenario in comparison with ICRP intervention levels; for comparison
the time after which the radiotoxicity in the disposed waste drops under the radiotoxicity
present in the fresh uranium required for the fuel production of fuel cycle A1 is also given
(Marivoet et al., 2009)

From a geological disposal point-of-view, it should be noted that the relevance of a scenario such as inadvertent human intrusion in an assessment of the post-closure consequences of a geological repository is contentious (National Academy of Sciences [NAS], 1995). For example, the considered geotechnical worker scenario assumes that, in spite of the fact that the drill-bit has to perforate a thick-wall metallic container, the geotechnical workers is not aware of the presence of very exotic materials in the underground and will not take precautions when examining the extracted cores. Recently, representatives of various radiological protection authorities discussed the treatment of human intrusion at an international workshop (NEA, 2010). They concluded that *"the results of human intrusion scenarios could also be used in a safety case to demonstrate the robustness of the repository system and the safety concept. The consequences to the intruder who comes into direct contact with the waste should not be required to meet regulatory protection goals. These consequences may need to be calculated but they should neither be evaluated against quantitative limit values nor be used as a crucial criterion for the repository optimization process."*

3.7 Other waste management aspects

3.7.1 Presentation of fast spectrum reactors as minor actinide burners

Fast reactors or ADSs will not burn all the minor actinides in one single cycle. Only a fraction will be "burned" and multi recycling of the actinides will be necessary to reach a significant reduction. Each cycle consists of cooling of the irradiated fuel, reprocessing, fuel fabrication and irradiation in a fast neutron facility. Dufek et al. (2006) estimated that it will take about 100 years to reach a substantial decrease of the actinides.

Present PWRs are capable of achieving burn-ups up to 60 GWd/tHM. Thereafter, build-up of fission products poisons the chain reaction and the reactor must be refuelled. Much higher burn-ups, in the range of 130 to 200 GWd/tHM, are considered for fast reactors and ADSs. The spent fuels that will arise from those high burn-ups will be much "hotter" (thermal output, gamma and neutron radiation, etc.) than the spent fuels presently considered for geological disposal. As a consequence, those fuels cannot be disposed of in the repository concepts that have been developed for disposal of spent fuel. In the advanced fuel cycles it is (tacitly) assumed that the spent fuels will be infinitely recycled. It has not yet been studied how such fuel cycles can be efficiently ended in the case of a nuclear phase out.

The long duration of the P&T process spread over several generations makes it a difficult issue for decision makers (Commission Nationale d'Evaluation [CNE], 2006; LaTourette et al., 2010). On such long timescales it cannot be excluded that alternatives for nuclear fission will become available such as nuclear fusion or alternative energy resources.

3.7.2 Management of volatile fission products

In the reprocessing plants at La Hague (France) and Sellafield (UK) the iodine released from the spent fuel is discharged, together with other volatile fission products, into the sea. On the other hand, the reprocessing plant at Rokkasho (Japan) captures the iodine on Ag-filters. However, a possible final waste matrix, which remains stable during several hundreds of thousands of years in disposal conditions, has not yet been developed. Continuing the discharge of iodine into the sea cannot be considered as a sustainable solution.

3.7.3 Generation of activation products

The inventory of the activation products was calculated within the Red-Impact project (Cunado, 2007), whereas most other P&T projects focus on actinide and fission product inventories. The generation of huge amounts of ^{14}C were calculated for the advanced fuel cycles. The large amounts of stainless steel used in the considered fast reactor design (i.e. European Fast Reactor) led to the generation of a considerable ^{14}C inventory. Nitride fuel was considered as the reference fuel in the ADS fuel cycle (B2); however, because of the large amounts of ^{14}C that were calculated to be generated, it was decided to also consider an oxide fuel as alternative fuel.

4. Conclusion

The introduction of advanced fuel cycles, based on fast neutron reactors, for electricity production can increase the efficiency of the use of natural uranium with more than a factor 100, because these reactors can burn all the uranium isotopes and not only the ^{235}U as it is the case in a PWR.

The volume of the high-level waste arising from advanced fuel cycles is about 5 times smaller than the volume of the spent fuel from the reference once-through fuel cycle. However, the dimensions of a geological repository are mainly imposed by the thermal output of the waste. Therefore, it is more important that the high-level waste arising from advanced fuel cycles generates about 3 times less heat than the spent fuel from the once-through fuel cycle. As a consequence, the needed length of the disposal galleries as well as the footprint of the geological repository is significantly smaller in the case of advanced fuel cycles.

The total doses due to geological disposal of spent fuel, vitrified high-level and intermediate-level waste in clay (or granite) formations are dominated by contributions of mobile fission and activation products. Geological repositories, in whose near field reducing chemical conditions prevail, very efficiently confine the actinides, because of their high sorption on clay minerals and low solubility. As a consequence the transmutation of actinides in fast reactors or ADS would have little impact on the doses due to the disposal of the radioactive waste in the case of the expected evolution scenario. On the other hand, in the case of altered evolution scenarios, e.g., in which the chemical conditions in the near field of the repository might become oxidising, some actinides can be dominant dose contributors.

The removal of ^{129}I upon reprocessing has a strong impact on the maximum dose. The development of very stable matrices for the immobilisation of iodine is highly desirable.

The radiotoxicity present in the high-level waste from advanced fuel cycles, after a few centuries, is drastically reduced by the transmutation of the actinides. This would lead to a considerable reduction of the potential hazard in the case of some disruptive scenarios such as an inadvertent human intrusion. However, the use of this low probability scenario as a decision making tool for long-term radioactive waste management is highly debatable.

Considerable amounts of mobile activation products, such as ^{14}C, are calculated to be generated in fast neutron facilities and, consequently, to give a significant contribution to the total dose; this observation confirms the importance of developing low activation fuel matrices and construction materials for future fast neutron facilities.

5. Acknowledgment

The authors would like to thank the European Commission for partially funding SCK•CEN's participation in the Red-Impact project as part of the 6th EURATOM Framework Programme for nuclear research and training activities (2002-2006) under contract FI6W-CT-2004-002408. We also thank our colleagues Lionel Boucher from CEA (Cadarache, France) for the neutronic calculations and Miguel Cunado from Enresa (Madrid, Spain) for the preparation of the waste inventories, which form the basic data for the presented analyses.

6. References

CNE (2006) *Rapport Global d'Evaluation*. Commission Nationale d'Evaluation, Paris

Cunado, M. (2007) *Final Report on Waste Composition and Waste Package Description for Each Selected Equilibrium Scenario*. Red-Impact Project, Deliverable 3.7

DoE (2008) Yucca Mountain Repository License Application: Safety Analysis Report. Chapter 2 *Repository Safety after Permanent Closure*. US Department of Energy, Washington

Dufek, J., Arzhanov, V. & Gudowski, W. (2006) *Nuclear Spent Fuel Management Scenarios: Status and Assessment Report*. SKB, Stockholm, report R-06-61

ICRP (1996) *Age-Dependent Doses to Members of the Public from Intake of Radionuclides: Part 5 Compilation of Ingestion and Inhalation Dose Coefficients*. ICRP Publication 72, Annals of the ICRP, 26(1)

ICRP (2000) *Radiation Protection Recommendations as Applied to the Disposal of Long-lived Solid Radioactive Waste*. ICRP Publication 81, *Annals of the ICRP*, 28(4)

LaTourette, T., Light, T., Knopman, D. & Bartis, J.T. (2010) *Managing Spent Nuclear Fuel: Strategy Alternatives and Policy Implications*. Rand Corporation, Santa Monica

Marivoet, J., Peiffer, F., Cunado, M. & Vokal, A. (2007) *Final Report on Waste Management and Disposal*. Red-Impact Project, Deliverable 4.3

Marivoet, J. & Weetjens, E. (2008) Impact of Advanced Fuel Cycles on Geological Disposal in a Clay Formation. *Nuclear Technology*, Vol. 163, No. 1, pp. 74-84

Marivoet, J., Cunado, M., Norris, S. & Weetjens, E. (2009) Impact of Advanced Fuel Cycle Scenarios on Geological Disposal. *Proceedings of Euradwaste '08: 7th EC Conference on the Management and Disposal of Radioactive Waste*, 20-22 October 2008, Luxembourg. EC, Luxembourg, report EUR 24040, pp. 141-151

Marivoet, J. & Weetjens, E. (2009) Impact of Advanced Fuel Cycles on Geological Disposal. *Proceedings of Scientific Basis for Nuclear Waste Management Symposium XXXIII*, Saint-Petersburg, 25-29 May 2009. Materials Research Society, Warrendale, Volume 1193, pp. 117-126

NAS (1995) *Technical Bases for Yucca Mountain Standards*. National Academy of Sciences, National Academy Press, Washington

NEA (2002) *Accelerator-driven Systems (ADS) and Fast Reactors (FR) in Advanced Nuclear Fuel Cycles: A Comparative Study*. Nuclear Energy Agency, Paris

NEA (2006) *Impact of Advanced Nuclear Fuel Cycle Options on Waste Management Policies*. Nuclear Energy Agency, Paris, report No. 5990

NEA (2008) *Nuclear Energy Outlook 2008*. Nuclear Energy Agency, Paris, report No. 6348

NEA (2010) *Towards Transparent, Proportionate and Deliverable Regulation for Geological Disposal: Main Findings from the RWMC Regulators' Forum Workshop*, Tokyo, 20-22 January 2009. Nuclear Energy Agency, Paris, report No. 6825

ONDRAF/NIRAS (2001) *SAFIR 2: Safety Assessment and Feasibility Interim Report.* ONDRAF/NIRAS, Brussels, Belgium, report NIROND 2001-06 E

POSIVA (2010) *Interim Summary Report of the Safety Case 2009.* POSIVA Oy, Olkiluoto, report 2010-02

SKB (2006) *Long-term Safety for KBS-3 Repositories at Forsmark and Laxemar – A First Evaluation. Main Report of the SR-Can Project.* SKB, Stockholm, report TR-06-09

Smith, G.M., Fearn, H.S., Delow, C.E., Lawson, G. & Davis, J.P. (1987) *Calculations of the Radiological Impact of Disposal of Unit Activity of Selected Radionuclides.* National Radiological Protection Board, Chilton, UK, report R205

von Lenza, W., Nabbi, R. & and Rossbach, M. (Eds.) (2008) *Red-Impact: Impact of Partitioning, Transmutation and Waste Reduction Technologies on the Final Nuclear Waste Disposal.* FZ Jülich, report vol. 15

Weetjens, E. & Marivoet, J. (2008) *Impact of advanced fuel cycles on radioactive waste disposal in a clay formation.* SCK•CEN, Mol, report ER-63

Permissions

The contributors of this book come from diverse backgrounds, making this book a truly international effort. This book will bring forth new frontiers with its revolutionizing research information and detailed analysis of the nascent developments around the world.

We would like to thank R. O. Abdel Rahman, for lending her expertise to make the book truly unique. She has played a crucial role in the development of this book. Without her invaluable contribution this book wouldn't have been possible. She has made vital efforts to compile up to date information on the varied aspects of this subject to make this book a valuable addition to the collection of many professionals and students.

This book was conceptualized with the vision of imparting up-to-date information and advanced data in this field. To ensure the same, a matchless editorial board was set up. Every individual on the board went through rigorous rounds of assessment to prove their worth. After which they invested a large part of their time researching and compiling the most relevant data for our readers. Conferences and sessions were held from time to time between the editorial board and the contributing authors to present the data in the most comprehensible form. The editorial team has worked tirelessly to provide valuable and valid information to help people across the globe.

Every chapter published in this book has been scrutinized by our experts. Their significance has been extensively debated. The topics covered herein carry significant findings which will fuel the growth of the discipline. They may even be implemented as practical applications or may be referred to as a beginning point for another development. Chapters in this book were first published by InTech; hereby published with permission under the Creative Commons Attribution License or equivalent.

The editorial board has been involved in producing this book since its inception. They have spent rigorous hours researching and exploring the diverse topics which have resulted in the successful publishing of this book. They have passed on their knowledge of decades through this book. To expedite this challenging task, the publisher supported the team at every step. A small team of assistant editors was also appointed to further simplify the editing procedure and attain best results for the readers.

Our editorial team has been hand-picked from every corner of the world. Their multi-ethnicity adds dynamic inputs to the discussions which result in innovative outcomes. These outcomes are then further discussed with the researchers and contributors who give their valuable feedback and opinion regarding the same. The feedback is then collaborated with the researches and they are edited in a comprehensive manner to aid the understanding of the subject.

Apart from the editorial board, the designing team has also invested a significant amount of their time in understanding the subject and creating the most relevant covers. They scrutinized every image to scout for the most suitable representation of the subject and create an appropriate cover for the book.

The publishing team has been involved in this book since its early stages. They were actively engaged in every process, be it collecting the data, connecting with the contributors or procuring relevant information. The team has been an ardent support to the editorial, designing and production team. Their endless efforts to recruit the best for this project, has resulted in the accomplishment of this book. They are a veteran in the field of academics and their pool of knowledge is as vast as their experience in printing. Their expertise and guidance has proved useful at every step. Their uncompromising quality standards have made this book an exceptional effort. Their encouragement from time to time has been an inspiration for everyone.

The publisher and the editorial board hope that this book will prove to be a valuable piece of knowledge for researchers, students, practitioners and scholars across the globe.

List of Contributors

David Bailly
IMFT, Institut de Mécanique des Fluides de Toulouse, France
IRSN, Institut de Radioprotection et de Sûreté Nucléaire, France

Christophe Nussbaum
SWISSTOPO, Mont Terri Consortium, Switzerland

Rachid Ababou and Hassane Fatmi
IMFT, Institut de Mécanique des Fluides de Toulouse, France

Jean-Michel Matray
IRSN, Institut de Radioprotection et de Sûreté Nucléaire, France

Donald M. Reeves, Rishi Parashar and Yong Zhang
Desert Research Institute, USA

Libby Clayton
BP Exploration, Anchorage, AK, USA

Shas V. Mattigod, Dawn M. Wellman, Chase C. Bovaird, Kent E. Parker and Kurtis P. Recknagle
Pacific Northwest National Laboratory, Richland, WA, USA

Marc I. Wood
CH2M Hill Plateau Remediation Company, Richland, WA, USA

V. V. Kostyuchenko, A. V. Akleyev, L. M. Peremyslova, I. Ya. Popova, N. N. Kazachonok and V. S. Melnikov
Urals Research Center for Radiation Medicine, Chelyabinsk, Russian Federation

Michal O. Schwartz
MathGeol, Germany

Constantin Marin
"Emil Racovit" Institute of Speleology of the Romanian Academy, Bucharest, Romania

Jan Marivoet and Eef Weetjens
Belgian Nuclear Research Centre (SCK•CEN), Belgium